BENEATH
THE
SKIN

A BODY OF ESSAYS

The pieces in *Beneath the Skin* were originally commissioned, performed and broadcast as part of BBC Radio 3's ongoing series *A Body of Essays*, originated and produced by Kate Bland at Cast Iron Radio.

wellcome
collection

WELLCOME COLLECTION is a free museum and library that aims to challenge how we think and feel about health. Inspired by the medical objects and curiosities collected by Henry Wellcome, it connects science, medicine, life and art. Wellcome Collection exhibitions, events and books explore a diverse range of subjects, including consciousness, forensic medicine, emotions, sexology, identity and death.

Wellcome Collection is part of Wellcome, a global charitable foundation that exists to improve health for everyone by helping great ideas to thrive, funding over 14,000 researchers and projects in more than seventy countries.

wellcomecollection.org

BENEATH THE SKIN

Great writers on the body

P

PROFILE BOOKS

wellcome collection

collection

183 Euston Road
London NW1 2BE
www.wellcomecollection.org

Copyrights in the individual works retained by the contributors

'Diality' from *I Knew the Bride* by Hugo Williams © Hugo
Williams and reproduced by permission of Faber & Faber Ltd.

The authors and publisher would like to make clear that some
names have been changed in the text to protect privacy.

1 3 5 7 9 10 8 6 4 2

Typeset in Granjon by MacGuru Ltd
Printed and bound in Great Britain by
CPI Group (UK) Ltd, Croydon CRO 4YY

The moral right of the authors has been asserted.

ISBN 978 1 78816 096 4
eISBN 978 1 78283 467 0

CONTENTS

INTRODUCTION

THOMAS LYNCH

'To have a body is to learn to grieve', wrote Michael Heffernan in his poem 'In Praise of It'. This was the opening line of the penultimate poem in his first of now many collection of poems – a book, like most slim volumes of verse, ignored on five continents, by an author who is internationally unknown, but who has, none the less, happened upon a truth, to wit: it is only in the body that our longings reside, our sorrows, our joys. If our hearts are broken, they are tucked beneath our sternums, snug in our pericardia, thumping out their iambic tunes. Mostly in the bones is where we ache for the embrace of another of our kind, or feel the residual of ancient wounds, old damages, wars long lost or won or fought to standstills. And only through the parts of bodies does mortality work its way into our demise – the cancer or the cardiac arrest, infarction, aneurysm or embolus. We are an incarnate species, embodied, brought into being by the conduct of other bodies, their parts and aspects, attachments and penetrations, the workings of their mysterious components and consortiums.

Even words, we claim in faith, become the flesh.

And whilst we are men and women of parts, we are also, in our own flesh, singular enterprises, solo endeavours. 'Three cubic feet of bone and blood and meat,' as Loudon Wainwright wrote for his son, Rufus, to sing into the new century, in his tune, 'One Man Guy'.

So the collection of essays assembled here goes some way towards making sense of the human condition by examining its particulars. What about the bowel or brain-box makes us who we are?

Was it the bad heart valve or the club foot, the cancerous bladder or the high cheekbones that shaped the rich internal course of our personal narrative? We can only guess. Our mother's eyes? Our father's hairline? The freckles, the feet, the heart failure? Who can know how we came to be the ones we are?

We have gathered here a little catalogue of usual suspects, the shared systems of the greater and some lesser animals: guts and lungs, gall bladder and skin – the innards and outer parts in hopes that by knowing the part we might better know the whole of our predicament and condition.

How is it that the parts engaged to send out a presidential tweet storm one day can enact Rachmaninoff's Second Piano Concerto the following evening? And while the heart becomes a ready metaphor for love and longing, grief and bereavement, the core of being, what case might we make for the symbolism of the pituitary gland? Or when 'guts' are the locale of valour, the cerebellum where a soul might reside, to what end, we might wonder, that first segment of the small intestine, the duodenum? Could it be to further our interest in etymology? The medieval Latin whence its name comes – duodeni, meaning 'in twelves' – refers to the fact that, where small intestines are concerned, size apparently matters: the breadth of twelve fingers approximates the length of the duodenum.

We are wholes and parts, one of a *kind*, and *one* of a kind. Still, here is where the part exposes something of substance about the whole, and why writers and readers, as much as medicos and anatomists, ought to be eager to understand these details of our bedevilments and being.

The father of modern essaying, Michel de Montaigne, in an effort to understand his kind, advised the test and measures method, believing, as he wrote in his masterful *Of repentance*, that 'Each man bears the entire form of man's estate.' High in the library of his solitarium, he studied his body, its senses and sounds, gases and appetites, longings, desire. So, in that spirit, here are some bits and pieces, small gravities, an effort to better understand humanity by looking at the human being, the sapient animal that is Man by meditating on its parts.

INTESTINES

NAOMI ALDERMAN

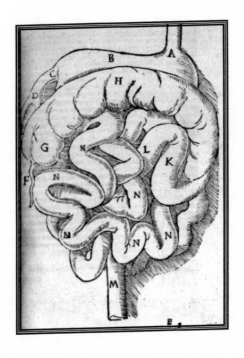

The proximity of the anus to the genitals, Freud tells us, is the source of much if not all human neurosis. It's fashionable to distance oneself from Freud these days, to say 'I wouldn't go that far' and 'of course Freud was sex-obsessed'. But I would go that far, and most humans are sex-obsessed.

The gut, frankly, is a problem. What it does is not only mysterious and puzzling – as are all our internal organs to a great extent – but also difficult for us to bear. And when we start to think about the symbolism of the gut, we might understand what Freud meant.

At one end of the gut is the mouth – a delightful place of many different kinds of joy. And then, at the other end, there's the anus. It produces farts, which stink of decay, foulness and poison. It makes poo, also foul-smelling, bearing disease, a sticky, stinky, brown contaminant. And it comes out of us! And not just out of our own bodies, but out of a hole *right next to* the parts of the body that can give us such great pleasure, whose development indicates adulthood, which can produce new life. It's like a terrible joke played by human biology, to drag us down from the heights to the depths, to remind us that whatever ecstasy we find, we're also, essentially and at all times, full of shit. This is why poo is so funny. This is why we have to laugh at it. If we didn't laugh, we'd cry.

For Ernest Becker, author of the Pulitzer Prize-winning book *The Denial of Death*, the anus and the shit it produces are more than just a joke – they're a terror. They represent the corruption of the flesh, the fate that awaits us all. 'What am I?' a child might ask herself. 'I am a thing which takes in beautiful, glowing, healthy, delicious, colourful, exciting food. And then what happens? I turn it into shit.' This is the inevitability of decay writ small, writ daily. It is the inevitability of death. 'The anus and its incomprehensible, repulsive product,' says Becker, 'represents not only physical determinism and boundness, but the fate as well of all that is physical: decay and death.'

The three-year-old daughter of a friend asked her mother what happens to the food she eats. 'Your body takes energy from it and then you turn it into poo,' replied the mother. Her daughter's crying was inconsolable. 'No mummy, no no,' she kept saying, 'no no no.' It is the same cry as in Julian Barnes' *Nothing to Be Frightened Of*, when he recounts his thanatophobia – fear of death – as he wakes in the night, 'alone, utterly alone, beating pillow with fist and shouting "Oh no, Oh No, OH NO" in an endless wail'. Shit is death. Death is serious. We have to laugh at shit. We can't take it seriously because it's so intensely serious.

The mouth, the anus, the intestines between them, transforming beauty to rot and deliciousness to disgust. It is in here that the rubber meets the road in our relationships with our bodies – where we come daily face to face with the corruption and decay that is our eventual inheritance. Bodies are mysterious, we are puzzles to ourselves. But it's here in the intestines that the obvious mystery is most apparent. If I can do this to food, what on earth am I?

When I was in my early twenties, my mother, then in her mid-fifties, was rushed to hospital with a burst intestine. The reasons for the rupture have never been particularly clear. Perhaps it was an infection in a fissure in the gut. Perhaps it was a weakness caused by the caesarean through which I was born, years earlier. Perhaps something else entirely. She had to have a colostomy bag for eighteen months, while her bowel healed; there's an experience to bring a family face to face with the realities of the workings of the gut. My mother's mother had also experienced some kind of ruptured bowel in her mid-fifties. I stare at my stomach and wonder what it has in store for me.

But that's not all. If the story of my family were written by a novelist, one might say that the symbolism around the stomach and the bowel and the process of digestion was just a little overdone, just a bit too obvious. A close relative was born with pyloric stenosis – one of the sphincters in his stomach wouldn't open – and his earliest days were long bouts of dramatic projectile vomiting while his mother tried to persuade the doctor that something was really wrong. He had to be operated on when he was only a few days old, just a tiny baby with a great long scar across his abdomen.

And these refusals of the stomach are only one side of the story. There are also the stomachs that are too welcoming, too efficient, too delighted to accept nutrition. I'm fat. My father is fat. My grandmother was fat. My aunt was fat, until she became a Weight Watchers leader. Our family stories are wound in a tight nexus around eating and not eating, digesting and not digesting, wondering how to get food to go through the body well, or how to stop it.

But I don't just think it's my family. Culturally, we are obsessed with food and diets. We create ever more luxurious constructions of fat and sugar – anyone fancy a cronut, an arrangement of buttery croissant pastry fried like a doughnut? But at the same time we come up with ever more punishing dietary regimes, everything from fasting two days a week to cutting up healthy stomachs because we find fatness, now, so culturally unacceptable. We watch celebrity chefs on television drizzling chocolate sauce or honey or butter, and associate food with sex from Jamie

Oliver's cheeky Naked Chef to Nigella's flirtatious looks and Gordon Ramsay stripping off his clothes at the start of *The F Word*. And at the same time eating disorders are on the rise, and our cultural beauty ideal becomes thinner and thinner, fuelled by Photoshop, when real human bodies don't look thin enough. Last year alone, there was an 8 per cent rise in young people being admitted to hospital with eating disorders in the UK.

We are worried about food, about digestion, about our stomachs. The intestine is the seat of our anxiety. And our anxieties are meaningful. To be anxious about something is to be obsessed with it. If you have constant anxious thoughts about a topic it's because, on some level, you are enjoying thinking about it. What is it about food and eating that gives such satisfaction in contemplation?

I suspect that it has something to do with Thanatos. It's been remarked before that the Victorians were obsessed with death, but couldn't bear to talk about sex, and we are just the other way around. Inflow, outflow. We talk about food and about youth and about sex. The starts of things. We live in those beginnings, as if it could be the first day of spring forever. If we keep on worrying and worrying about food – am I eating enough, or too much, is it the right sort – we can just flush our shit away in a clean tide of water and never think about it, or what it represents, again. If we focus on youth, we can send our elderly to nursing homes and not have to look at them or think about them. If we're always talking about sex, the beginning of everything, there won't be any room for death: the end of it all.

Is it possible, therefore, to be delighted by shit? And might we do better as a society and as individuals if we worked out how? I suspect it is, and I suspect that a full appreciation of the workings of our own horrifying poo machine, the intestine, could lead us in good directions in this enterprise.

Of course, shit can be delightful, as anyone who's ever suffered from constipation will affirm. My brother and his wife recently had a baby girl, making me an aunt for the first time. They've been so thrilled, we've all been thrilled, when she's done a long and healthy poo. Pooing means everything's working right. Inflow, outflow. Pooing means: this is how it's supposed to go. Death, at least when it comes at the end of a long and useful life, means the same thing of course. It might be that nature knows what it's doing; it might be that the process of decay, utterly out of our hands, has some beauty to it.

Contemplating the wonderful things that 'nature' knows, and that we have no idea about, might be a good point to introduce the neurons in your stomach and the nature of the mass of bacteria living in your gut right now. Did you know that you have brain cells in your stomach? They cover the walls of your gut. You have as many neurons in your digestive tract as a cat has in its head. Think of all the things a cat knows: what's nice and what's nasty, who to trust and who to steer clear of, where good food comes from and how to hunt it down. That's the kind of thing your stomach might know. No wonder we talk about a 'gut instinct'.

The neurons in the gut are connected directly to the brain via the vagus nerve, which enters your brain right next to the parts that deal with emotions. The stomach can seem to know things that we don't know ourselves. Experiments have been done in which people are fed food through a tube – they cannot taste or smell or chew it, but the entry of their favourite foods into their stomachs makes them predictably happier than some equally nutritious slurry. Your stomach knows things. You get butterflies in your stomach because those neurons down there have some idea about what's going on.

There are parts of 'ourselves' – perhaps the great majority of 'ourselves' – which we cannot access. In her memoir *The Shaking Woman*, Siri Hustvedt writes of a sense of duality she experiences when in the grip of a shaking fit. She has 'a powerful sense of an "I" and an uncontrollable other'. Our bodies, full of intelligence, our stomachs, full of neurons, are in some sense other 'selves' within us, communicating with the brain but not fully part of it.

There's an even more real 'uncontrollable other' within the gut, though. We think of ourselves as single, unitary, contained within this envelope of flesh; everything inside the outline of our skin is 'us'. And yet. Your gut contains a 'microbiome' – an ecological community of micro-organisms. They're the 'good bacteria' so beloved of adverts for probiotic yogurt. The cells of our gut flora are much smaller than the cells of our own tissue – so much so that 'we' actually contain *more* gut flora cells than human body cells. If I were to hold a referendum inside my skin with each cell getting a vote, 'I' wouldn't come close to taking office.

And that analogy isn't as ridiculous as it sounds. Gut flora can influence mood and health – everything from depression to rheumatoid arthritis can be improved by increasing the variety of flora in the gut (our guts, apparently, would like to be a proportional-representation government, the more variety the better). Our gut flora can release hormones which encourage us to eat more of the food *they* like. Moreover, we've only been able to culture about 5 *per cent* of the flora in the gut. We have no idea what the other 95 *per cent* are. All you can get in your probiotic drinks are that measly 5 per cent – for the rest, you'll have to wait until we've gene-sequenced the missing gut flora, a process which is ongoing. Or, if in dire straits, you might consider a faecal transplant, which is exactly what you think it is. Miracle cures have been achieved by inserting the poo of one person with a 'golden stool' into the gut of another, via a drip or a faecal enema. The new colonies of bacteria grow, and the recipients of the transplant start to feel better – it's worked on a range of conditions including rheumatoid arthritis and killer bacteria *C. difficile*. But don't try this at home.

The point is that what's going on inside our intestines is mysterious, astonishing, far more complex and far more intelligent than we imagine when we look at our stinky stool and think 'how did *that* come out of *me*?' The labyrinthine and beautiful arrangement of intestine sitting in the centres of our bodies has a brain, and our internal neighbours have desires.

And this is comforting regarding our major issue, Thanatos, too. I do not know how to digest food, but my intestine's got it covered – as well as some thoughts on how anxious certain situations and people make it feel. I may not know how to die, but my body's got it covered.

The French thinker Montaigne, the inventor of this very form, the essay, took a bad tumble from a horse and almost died of the resulting injuries. While his friends were horrified to see him clawing at his clothes and apparently in agony, he experienced a blissful, easy sensation. When he recovered, he wrote of his brush with death: 'if you don't know how to die, don't worry; Nature will tell you what to do on the spot, fully and adequately. She will do this job perfectly for you; don't bother your head about it.'

From our cultural food neuroses we can learn that

we're obsessed with beginnings, not endings. That we are troubled by the apparent limitlessness of our own desires – stoked by consumer capitalism. And that, although we know we'll eventually turn everything to shit, we don't want to think about it. But perhaps what we need is that thing so little discussed in modern Western society – a little faith. We may not know how to make our poo, but our stomachs do. We might not understand how to die, but our bodies will get us through. We know more than we think. And 'we' don't have to know it, to know it.

SKIN

CHRISTINA PATTERSON

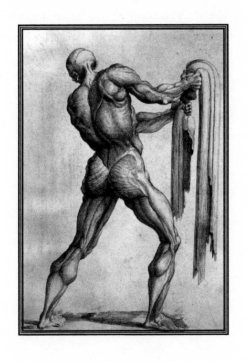

'It is,' said my father, 'like a peach.' He had just been stroking my cheek. It was the first time I gave any thought at all to the membrane separating me from the world.

I knew that it felt nice to lick ice cream off fingers, and to tip toe on sand and feel the slap of water against shins. I knew that when I shot too fast down the slide at nursery school, and landed with my face against a cold metal edge, there was suddenly a gash in a place that had been smooth. When my mother took the plaster off, she gasped. She said she hoped I wouldn't be scarred, but I was, and I am. Even when this happened, and when I watched knees slashed by barbed wire grow a brown crust like the shell of a beetle, and elbows bashed against bricks turn from white to speckled purple, I still never thought about this thing that marked the line from out to in.

Later, there were things called verrucas, which you got from going swimming, and which meant you had to go to a special clinic where they tried to burn them off. There were the welts you got when you brushed against nettles, or against the wrong kind of jellyfish when you tried to catch a wave. There were so many things that could scrape, or slash, or sting. But it wasn't until my father stroked my cheek, and talked about peaches, that I realised I had this thing my father thought was beautiful, called a skin.

I didn't know then that the skin of a child is different to the skin of a woman or a man. I didn't know that it was softer, and smoother, and nicer to touch, because there is more adipose tissue under it, and because the outer layer is still thick. I didn't understand that what is soft, and smooth, and sprinkled with a down of fine hairs, can make an adult feel happy and sad. It can make the adult's heart leap with joy, and the urge to protect, but it can also grip that heart with fear. When you're a child, you don't know that there's a thing adults call 'innocence', and that life will make sure it doesn't last.

When you're a child, you don't understand that fresh, young skin is seen as beautiful, because youth is seen as beautiful, and what is beautiful is to be prized. But you do learn that what's ugly isn't. You might, for example, hear stories about lepers from the Bible. You might hear about the man hailed as a messiah who touched a leper and told him to 'be clean'. He told him to 'be clean' because if you have this disease, which makes your skin scale, and your fingers and toes change shape, people think you're dirty. Lepers, you learn, had to live apart from everyone else, and sometimes they had to ring a bell to warn people that they were coming.

If you did hear stories from the Bible, you probably also learnt about Job. Job, the Bible said, was tested by God. God allowed Satan to kill Job's cattle, and his camels, and his donkeys, and his sheep. God allowed Satan to kill Job's sons and daughters, and he also allowed him to 'afflict' him with boils. Job's boils were so disfiguring that his three best friends didn't recognise him. When they did, they were so shocked they didn't speak for a week.

From the Bible, you learn that a skin condition is something that ought to make you feel ashamed. And then, just when you're beginning to think about how much you want to touch someone else's skin, and how much you want to feel their lips on yours, at a time, in fact, when the hormones coursing through your veins are making you think that what you want more than anything is to press your naked body against someone else's naked body, you start peering in the mirror and seeing spots.

If you're lucky, it's just a sprinkling of pimples, though even pimples can dent the fragile confidence of someone who's hovering between being a child and leaving childhood behind. Just when you're worrying that all your friends are gorgeous, and that you're too fat, or too skinny, or too tall, or too short, you can find your face peppered with tiny buttons of pus. People talk about this as if it's funny. Books, films and TV programmes seem to think teenage spots *are* funny. It doesn't seem quite so funny when you feel so ugly you don't want to leave the house.

And when the teenage years pass, and the spots don't, that doesn't feel funny at all. I know, because this is what happened to me. The spots, for me, were bad enough, but when I was 23 what happened on my face was more like a war. What happened on my face was so bad that when I went to a hospital for skin diseases, the consultant invited a group of students in to gawp. The doctor prescribed a treatment called PUVA. This meant I had to go to the hospital every day, and be blasted with a special kind of ultraviolet light in a metal box like an upright coffin. After a few weeks, the light had burnt off most of the spots, and several layers of skin. It didn't burn off the scars.

When my face was raging, and weeping, and pulsing with deep red lumps that were just waiting to grow their big, yellow heads, and peering in car wing mirrors and feeling sick at what I saw reflected back, I thought this was as bad as it could get. I now know that the acne I had, which was severe enough to get me referred to the top acne specialist in the country, was just dipping a toe in the water of what can happen to the skin.

At the Gordon Museum, at Guy's Hospital, for example, you'll find faces, arms and legs which don't look much like faces, arms and legs, because they're covered in scales, or lumps, or bumps, or swollen with big, fleshy growths like horns. Some of them *are* horns. 'Cutaneous horns', according to the curator. They are, he told me matter-of-factly, made of 'soft tissue'. In glass jars, you'll see carcinomas and melanomas, and human skin that looks like the skin of a lizard. In one glass jar, you'll see the head of a woman. She is old, but has bright red hair, though what's more shocking than the hair is the giant, scaly growth that swells out of her brow.

If you thought too much about the stories preserved within the museum's collection, you might go mad – if, for example, you thought of the man whose giant, swollen foot looked like the foot of something that wasn't human. 'Trench foot,' said the curator; but in all those stories and poems from the First World War, trench foot at least made you think of something that would look like a foot. Or if you thought about the Chinese patients who had their portraits painted before their operations, whose giant tumours looked like extra shoulders or backs; they had them cut off, without anaesthetic. Miraculously, they all survived.

And then there's the baby. It's called a 'harlequin baby'. The word 'harlequin' makes you think of a clown. But when you see the baby, curled up in its jar, comedy is the last thing on your mind. When you see the diamond scales that cover it, where the skin has cracked and split, you think of the mother who carried this baby, and of what she felt when she held it, briefly, in her arms.

When I went from dermatologist to dermatologist, and tried lotion after lotion, and all the drugs that they could think of, I didn't know about all the terrible things that could happen to the skin. But I did learn quite a lot about how the skin works. I bought books with titles like *The Acne Cure* and *Super Skin*. I also had a book called *Acne: Advice on Clearing your Skin*. 'Acne,' it said, in the first line of the first chapter, 'is a skin disease that we still need to research'. It is, in other words, a disease that doesn't yet have a cure. The book had diagrams showing the 'tough outer layer' of skin called the 'stratum corneum', which is, it said, like a 'protective coat'; the 'epidermis' under it, which produces cells which move up to the 'horny layer', and the 'dermis' underneath it, containing blood vessels and nerves. 'It takes,' said the book, 'about 28 days for an epidermal cell to travel from the base of the epidermis to become a horn cell at the top of the epidermis'. It takes, in other words, twenty-eight days for the skin to renew itself. New skin, new you, in less than a month.

The trouble is, most skin conditions don't go away in a month. My dermatologist founded an acne support group, and you don't run support groups for things that are gone in a month. Another book I bought was called *Learning to Live with Skin Disorders*. It didn't say that what you should do, if your skin is flaking, and scaling, and peeling off, is open a packet and take a pill. For many people, skin conditions are for life. 'Why did I marry so young?' said the novelist John Updike, who had psoriasis from the age of six. 'Because, having once found a comely female who forgave me my skin, I dared not risk losing her and trying to find another.' He wrote this in a book called *Self-Consciousness*, in an essay called 'At War With My Skin'.

But the science is advancing. At the Centre for Stem Cells and Regenerative Medicine, for example, biologists are looking at the way our skin cells respond to the environment, and the role stem cells can play. The skin, the epidermis, the hair follicles and the sebaceous glands all have their own stem cells. If you have a wound, stem cells start doing things they wouldn't usually do. Stem cells, in fact, might be our salvation. And which of us with a skin condition hasn't wanted to be saved?

I don't know what won the war with my skin. In the end, it could have been time. But what I've learnt, in all my years of failed medical treatments, and failed trips to homeopaths, and naturopaths, and acupuncturists, and herbalists, is that the skin often speaks when we can't. When we are sad, and angry, and lost, and lonely, our skin bubbles, and itches, and weeps. We can pop the pills, and slather on the potions, but putting a muffle over someone's mouth only stops the words from getting heard. Half the time, we don't know what they are. Maybe most of the time we don't know what they are. All we know is that something about how we're living our life – our work, our family, our home, our pysche – is making our skin crawl.

If you doubt that the mind can have a physical effect on your skin, then have a look at the research. In a Japanese study of contact dermatitis, all the people taking part were touched with harmless leaves but told they were touched with leaves which have the same effect as poison ivy. They all reacted to the leaves that ought to do no harm. And there are many studies which show that people get rashes when people they love die. 'Skin disorders,' says the psychoanalyst Darian Leader, 'are often symbolic, yet involve tissue change'. In his book *Why Do People Get Ill?*, he describes the case of a young soldier who developed a rash that looked like the welts you might get from a whipping. He got the whipping when he had peered through the window of a girls' dormitory when he was nine. He got the rash ten years later, when he was found loitering outside the nurses' dormitory at his military post. He was hoping to see a particular nurse, but was stopped by an officer and told off. He had the welt-shaped rash within an hour.

The skin is designed to protect us from the world. No wonder it often doesn't feel thick enough. We need, we say, to grow an extra skin. We want, we say, to be comfortable in ours. What we don't want is for our sadness, or our fear, to be written on our face.

The miracle, perhaps, is this: that most of the time, it isn't. Most of the time, for most of us, this organ, the biggest of the body, isn't covered in rashes or weeping sores. Most of the time, for most of us, our skin does what it's designed to do. It holds everything together. It keeps your body at the right temperature to survive. It stretches and shrinks as it needs to; it protects you from danger and it alerts you to pain. It also allows you to feel the warmth of sunlight, and the electric joy of a lover's touch.

As the sun rises and sets, as the moon waxes and wanes, and as the seasons change, the skin pumps out new cells. Whatever happens in our lives, it keeps pumping out those cells. When you have a wound, it heals. There might be a scar, but it does heal. What's left might not look like a peach. When you've lived a bit, your skin doesn't look like a peach. When you've lived a bit, that elastic barrier between you and the world shows some of the battles you've fought, and won. We should see the beauty in those scars.

NOSE

A. L. KENNEDY

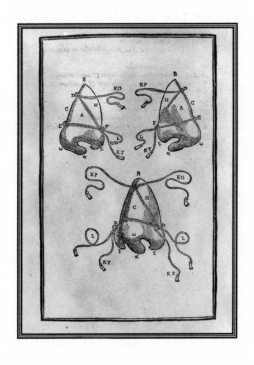

In Nikolai Gogol's story 'The Nose', a civil servant called Kovalyov wakes one morning to discover his nose is missing; in its place there's only a smooth, flat space. Without a nose, Kovalyov finds he can't work, can't eat, he's scared even to go outside. As for his girlfriends – well he discovers, noselessness, apart from being a kind of facelessness, seems to imply other, lower, deficits. Worse still, freed from Kovalyov, his nose is swanning around St Petersburg dressed for success in a 'gold-braided, high-collared uniform, buckskin breeches, and cockaded hat'.

Gogol himself had a famously generous nose – but this tale is less a personal statement and more an absurdist satire on Tsarist Russia's obsession with rank. Still, there's plenty of nose-related wisdom here. Humans may be hard-wired to love and protect infantile faces – ones with big eyes, big foreheads and not much nose – but we still all have multiple reasons to treasure our noses. They go bravely before us throughout our lives, gently drooping and appearing to grow as time passes, perhaps to indicate increasing maturity and resourcefulness. The nose helps to form our expressions and our faces seem so strange to us if we do lose or damage them that we have a long history of fashioning cosmetic replacements. Sixteenth-century astronomer Tycho Brahe had a brass one. British troops disfigured in World War I were supplied with carefully painted tin noses and rudimentary skin grafts. The earliest recorded graft – replacing the tip of a nose – was achieved in India around 1795. Plastic surgery can now do a great deal to create or reconstruct our noses. The immense popularity of elective rhinoplasties – 'nose jobs' – indicates how important this most public organ's perceived perfection is to our self-esteem.

The sense of smell our nose enables conjures memories faster than conscious thought and puts flavour in our meals. Much of what we taste actually comes from scent. Try eating an apple while, for example, smelling petrol if you don't believe me. Patients who develop anosmia, an inability to smell caused by accident or disease, will generally report diminished appetite and somewhat joyless meals. Smell can change your mind. Studies have shown that inhaling the scent of garbage can affect moral judgements and make you more politically conservative. Being forced to smell warmed vanilla by a wily estate agent as you view a house may make you think – 'Mm, I *must* buy this apartment – it smells of cake and happy childhoods.' Each inhalation gives us breath to speak, sing, swear, live. Our sensitive olfactory bulbs can register chemical upon chemical, letting our brains enjoy anything from the isomers of rose oxide in a bouquet to the hundreds of compounds in the smell of coffee.

We actually have four nostrils, the two external and two internal right at the back of the nasal cavity by the entrance to the throat. The nostrils oscillate their action to help make sense of complicated scents and their location. The external nostrils are equipped with something like a thousand hairs apiece. These are vibrissae – they're what used to be our whiskers. They help us cleanse each inhalation, as does mucus. Propelled by microscopic cilia on the surface of the cells lining the nose, our mucus contains chemicals to fight disease and resist pollen. And, all day, every day our noses humidify up to 14,000 litres of air so we can breathe it more efficiently and comfortably. Kovalyov was right to fear that leaving home without a nose would be risky.

Out in the world I rely on my nose to prevent social disaster. I have pathetic facial recognition, but even a brief encounter with someone's scent will fix them in my memory for years. But repeatedly explaining my disability has revealed to me that any reference to smell can be socially disastrous in itself. Smell is personal, animal, basic. Simply mentioning it can provoke uneasy laughter, if not hysteria. And it's not unusual to find our complex, helpful, wonderful noses reduced to a punch line.

We laugh at noses. The red nose is the only part of a

clown that isn't scary. Even without the rest of the outfit it can add instant jollity. This may be a formalised way of mocking the noses, purplish with broken capillaries that we associate with habitual drunkenness, homeless tramping, or hard outdoor labour. When clowns seem threatening, perhaps it's because they're effectively the unruly poor coming to get us.

The Marx brothers were immensely talented comedians, but their imposing noses gave them a head start. Joke shops still sell Groucho glasses – or beaglepusses – with the plastic nose attached. An iconic nose has outlived its owner. Einstein was a genius and a great communicator – but is his lasting reputation all down to the enthralling attractions of theoretical physics? Don't we, just a little, find his mankind-dwarfing, imagination-bending concepts warm and memorable because of his amiable, noticeable nose? Cyrano de Bergerac, obscure author of the world's first science fiction novels, had, as far as we know, an above-average nose, forging ahead through duels, debates and flights of fancy. Edmond Rostand's eponymous drama enlarges Cyrano's nose and creates an unforgettable hero. Cyrano's drama is unspeakably tragic, in part to stop us laughing too much at that nose – laughing as we do when we see Jimmy Durante sing a tender love song, or Woody Allen holding a gun to all that's left of a dictator – the nose – in his futuristic comedy *Sleeper*.

And we don't just mock noses – we can seem to hate

them. Their prominence apparently offends us. Inappropriate curiosity *pokes them into things*. Steamy TV medical dramas capitalise on actors' alluring eyes by covering their unromantic noses with surgical masks. The cliché of enticingly veiled female beauty operates in similar territory. The nose is what we look down, or turn up. Or else we simply, dumbly follow it.

While our first and simplest words for smell are linked with animal intimacies (our mother's skin, her hair), they're more often for smells that are viscerally unpleasant – and quite possibly our fault. Sigmund Freud thought that smell was primitive, inextricably linked with the anal stage of development. Even neutral words for the quality of having scent – smelling – aren't that neutral. As you'll discover if you tell your beloved 'Darling, you smell …' Even if the end of that sentence is 'of sweetshops and paradise', its start may have already shattered a budding relationship. Humans are animals but don't want to smell like them. Billion-dollar industries exist to save us from body odour, foot odour, bad breath, sweat. Before we knew about microbes, we even blamed infections on bad smells – 'miasmas'. And if neutral words for smell are severely limited, words for bad smells are apparently unlimited: stink, stench, reek, pong, honk, howff, hum, ming – das stinkt, eso apesta, ça pue, это воняет.

There's a neurological reason for our bias. Smells associated with disgust take a shortcut through the amygdala

– part of the brain's very emotional, unnuanced limbic system. They reach us at a basic, animal level. More pleasant or neutral smells are processed in our cortex – the clever, fancily evolved layer that lets us invent string cheese and deodorant and rise above emotional involvement with aromas. Evolutionarily speaking, bad smells are about danger, decomposition, fear, pain, fleeing, fighting – it's important to be able to detect them and react swiftly. When we talk about something morally disgusting we may say it has a bad smell, it stinks, and that's a clue that our brains process metaphorical disgust the same way they process the real kind. So disturbing smells get VIP attention in case they're going to kill us – but everything else? Smell is so important to survival that it has many connections to parts of the brain that evolved early, like the limbic system and the brain stem. Perhaps we treat it like an unwelcome intruder because it operates deep within us, beneath our control over our thinking. And smell has relatively few connections to the left neocortex, where we keep our words. This means our power to describe smells that aren't potentially fatal is pretty stunted. The orchestral blending of complex scents in a forest at dawn is ... *nice? Countryish? Foresty?* Chocolate smells – *of chocolate*. Smells don't have a separate vocabulary, even for scent sophisticates; noses who make a living out of categorising wines or perfumes will describe scents and flavours in terms of other things: hints of sandalwood and eggshell, an aftertaste of tarmac and so forth. We can pick out sharpness, sweetness, acridity and not that much beyond. Only a few smell-sensitive cultures – often developed in light-deprived environments – have a range of other smell words. Tribes in the Andaman Islands, Papua New Guinea and the Amazon have terms for subtly connected smell groups. For them a scent can clearly resemble others in a scent group, the way that blue sky, a blue police box and baby blue are all different, but still identifiably blue. Some researchers think smell-centredness may be an early trait from our Denisovan ancestors, still carried in some humans' DNA. I myself long for a world where smells are allowed to form palettes, families, proud dictionaries of aroma. Many languages do have one word for a complex smell so universal and useful it has stayed with us, perhaps from our days as hunter-gatherers. In English the term is *petrichor* – the smell that lets us know when it's going to rain.

It has been suggested by some academics, of course, that peoples prioritising a 'primitive' sense may be especially primitive, but that could be our anti-smell bias talking. Sniffing, scenting, allowing an inrush of information that speaks to us without words – it's not appealing to everyone. It can seem more suited to dogs or at least hairier primates than Homo sapiens. Throughout the ages those with money and power have made sure not to smell like the poor and not to build houses downwind of them.

Civilisation has consistently been associated with not smelling, or certainly not smelling natural. The doggedly cerebral and moderate Plato even thought use of perfumes led to effeminacy and depravity. Even Kant took against smell. The nose has been associated with all that's indecorous, dirty – even helplessly sexy – and we've laughed at it in revenge.

We should, of course, have thanked it. When early-twentieth-century neuroscientists were trying to understand brain structures, they dissected rats and noted the massive olfactory bulbs with which the rats had, until recently, been trying to understand the neuroscientists. Intimately linked to these bulbs was an area of the rat brain initially christened the rhinencephalon – the nose brain, what we now call the limbic system in rats and humans. It's not just associated with tripping alarms, states of arousal and processing emotion; it helps create our memories. That's why certain scents aren't just animal invaders: they're time travel, they're joy, they're home, they're heartbreak.

I'll never forget, years after his death, being passed on the street by a man wearing my grandfather's aftershave. For one deep moment I could summon that voice, that face, being folded in those arms again. A gift of the nose.

Although, yes, I'll admit some of the gifts can seem unsettling. It's okay that Mrs Rat can know by scent that now is the time to make rat babies with that male rat over there, or can recognise her relatives, including those rat babies. Mrs Rat may even coordinate her reproductive cycle with her close lady rat neighbours, because they've inhaled each other's pheromones. We humans (even me) rely most on vision – the cool, sophisticated fashion photographer among the senses. But humans recognise relatives and choose mates by scent. They can detect fertility and even specific genes. We may find a face more attractive depending on how it smells, and we tend to choose perfumes that enhance our own natural blend of scents. We spend a fortune trying to obliterate our natural smells, but pheromones still change our moods, our focus levels and how we perceive each other, and can synchronise women's menstrual cycles. For centuries phallic masks and racy nose-related humour helped us deal with that weirdly sexy something about the nose. We now know scent prepares us for intimacy and maintains it – the nose even contains erectile tissue… And the people with whom we mate, form relationships? We love their scents, up close; our bodies continue each other in each breath. No wonder the artfully wild and passionate Romantic Movement embraced scent.

Our noses give us breath, life: the perfume of our children's skin, or of our lover's tenderness, the hallway scent of being home, the pleasure in every bite, the power to turn back time. So no more jokes, or shaming, let us carry our noses ahead of us with pride.

APPENDIX

NED BEAUMAN

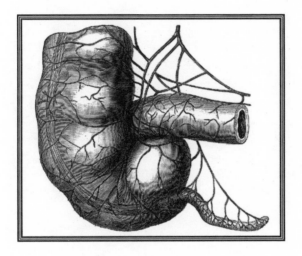

Over the past few years I've spent a lot of time in New York, in theory to promote the American editions of my novels, in practice to eat tacos and drink bourbon. Many of the people I've met there are freelancers or artists of some description, and they tend to be pretty alarmed about the prospect of President Trump getting rid of the Affordable Care Act, better known as Obamacare, to replace it with 'something terrific' of his own design. Americans with nine-to-five jobs get their health insurance through their employers. Before Obamacare, if you didn't have an employer, and you didn't have a union, and you weren't eligible for any government programs – if you were a struggling novelist in Brooklyn, say – insurance was probably out of your reach. The American healthcare system is the most expensive in the world, and gets about 5 per cent more expensive every year. This means that if you come down with something serious, need hospital treatment and don't have insurance, the bills can leave you destitute.

Because I'm not a permanent resident of the US, I'm not eligible for Obamacare, so when I'm over there as a tourist I take out travel insurance. But I've never really trusted travel insurance, because if you call your provider it's almost impossible to get them to confirm in advance whether you'd be covered for a given mishap. In an emergency, you just have to pay all your own expenses up front and hope to God there isn't some eerily specific clause in the fine print that absolves them from reimbursing you afterwards. As a British taxpayer, I take some comfort in the knowledge that if I got a chronic illness like cancer, I could fly straight back to London and collapse into the welcoming arms of the NHS. But of course sometimes you can't just get on a plane. Whenever I spend time in New York, I have two neurotic fears at the back of my mind. One is that I'll be hit by a taxi. The other is that my appendix will burst.

Before I started researching this essay, I held certain firm beliefs about the appendix, as follows. It is a vestigial and functionless part of the body, like wisdom teeth or goosebumps, or my scaly prehensile tail. That, at any moment, unprovoked, without warning, it can burst, whereupon you feel an agonising pain in your side and have to go straight to hospital. And, if it's a hospital in the US, you will wake up after the operation and inside the perfumed envelope on the silver platter beside your bed will be a bill for a hundred thousand dollars.

I have since learned that none of those things are true. I'll go through them in reverse order. According to one estimate, the average cost of an appendectomy in the US is only $14,000, or about £10,000. That's still a lot of money, but not nearly as much as I feared. I have no idea where I got that hundred thousand dollar figure from. Also, supposedly, travel insurance does cover the operation. Maybe one day I'll find out for sure one way or the other, although, like most of the practical wisdom we accumulate as adults, this information will arrive just slightly too late to be useful, since it's not like I'm ever going to get a second appendectomy – unless it turns out that I'm some sort of medical marvel, and I'm able to regrow my appendix, or I've been born with appendix triplets.

One of my favourite films is *The Wages of Fear*, Henri Clouzot's masterpiece of 1953, in which four desperate men drive trucks loaded with nitroglycerine over South American mountain roads, knowing that they only have about a 50 per cent chance of survival because their cargo could explode at any time. This is how I used to regard my appendix. I now know that *The Wages of Fear* does not, in fact, offer a very exact parallel. The appendix, a small pouch resembling a sort of boneless finger, hangs off the end of your large intestine on the lower right-hand side of your body, and it doesn't just go bang. Appendicitis, which is a gradual inflammation, is a different thing from appendiceal rupture, which is when your appendix actually starts leaking poison into your gut. Long before your appendix pops, you should feel it begin to swell, giving you plenty of time to go to the hospital for treatment. In some cases you don't even need an operation, just antibiotics, although in other cases you really do need the appendix to be surgically deported before it kills you with sepsis. Perhaps everyone else already knew this, but I didn't. A better parallel for the appendix is virginity. As a teenage boy, I used to feel that unless my virginity was removed as soon as possible it was just going to get bigger and bigger until there was no room in my body for anything else and I would die in agony. This may offer a Freudian basis for my deep-seated envy of Americans with proper health insurance, who can have a crucial excision performed at the drop of a hat, instead of belatedly and perhaps at considerable personal cost.

The third of my misconceptions about the appendix was that it's an obsolete technology – that, to borrow a relevant expression from the American south – it's as useless as tits on a boar. Most of us are taught at school that the appendix has no function in humans. Even medical students are taught this. One textbook from the 1980s asserts that 'its major importance would appear to be financial support of the surgical profession'.

But that's just not true. When I was tiptoeing around the Lower East Side, feeling like I had nitroglycerine in my lower torso, it never would have occurred to me that one day I'd be defending the appendix, but the fact is the organ has been slandered. The pervasiveness of this slander may be something to do with our understandable deference to the authority of Charles Darwin. In *The Descent of Man* he argued, influentially, that the vermiform appendage of the caecum was a 'rudiment', an organ 'bearing the plain stamp of inutility', which we were stuck with only because our forebears were lower mammals who might have needed it for digesting leaves and grass. Today, evolutionists still bring up the appendix in arguments with creationists, since if God really did design the human body from scratch it's quite hard to explain why he would have left room for this little bastard who does nothing but sit around all day and plot how to ruin you with hospital bills.

However, within twenty years of Darwin's death, doctors had noticed that the appendix was replete with lymphatic tissue. One Scottish anatomist, Dr Richard J. A. Berry, used this as evidence to assert quite firmly that 'the vermiform appendix of Man is not … a vestigial structure. On the contrary, it is a specialised part of the alimentary canal.' And this was in 1900, a very long time before any of us were being misled at school. Because lymphatic tissue is important to our immune system, subsequent biologists speculated that the appendix might have some immune function. Then, in 2007, Dr William Parker and a team of researchers at Duke University in North Carolina published a paper that may have finally vindicated the appendix.

These days we all know from yoghurt adverts that a

lot of the bacteria in the body are benign or even essential to our health. Each of us carries about a hundred trillion microorganisms in our intestines; this population is called the microbiome. If you ever get an infection that causes severe diarrhoea, what is happening is that all of the bacteria in your gut is getting flushed out of your body as a sort of nuclear option. Afterwards, you've shaken off the bad bacteria, but you've also expelled the good bacteria. Parker suggests that the appendix functions as a Noah's Ark, from which the good bacteria can be repopulated after the floodwaters recede. That's what explains all the lymphatic tissue: in the appendix, the immune system maintains structures called biofilms, which are safehouses for bacteria.

If Parker is right, then people who still have their appendices should recover much more quickly from serious gut infections than people who've had them removed. This isn't easy to test, because not many people in the developed world get cholera any more. However, a recent study at a New York hospital looked at patients with Clostridium difficile, which is a nasty infection you can pick up if your gut flora have been scoured away by a course of antibiotics. The study found that people without their appendices were four times more likely to have a recurrence of C. difficile. According to Parker's model, that's because those people had no way of rebuilding their microbiomes fast enough to defend against the next attack.

When I spoke to Dr Parker, he told me another fascinating thing about the appendix. Until around Darwin's time, appendicitis was rare or non-existent. Medical writings from ancient Greek and Roman times don't mention anyone dying of anything that we can now recognise as appendicitis. Even today, preindustrial societies in Africa and South America don't have to worry about the affliction. We can't know for certain why this is, but Dr Parker argues that appendicitis is a modern dysfunction of the immune system just like asthma and allergies, which people also don't get as much in the developing world. In the developed world there's plenty of hot water and soap to go around, and we live in relatively sterile environments. As a result, our immune systems, to borrow a metaphor from Dr Parker, become like bored teenagers: they have nothing to do so they find something stupid to get up to, which in this case is causing an inflammation in your appendix for no good reason.

This means that the appendix has an entirely different status in the developing world, for two reasons, both to do with hygiene. Firstly, you're much more likely to get a gut infection there, which means your appendix is much more likely to come in useful when you need your bacteria to be repopulated afterwards. Secondly, your immune system is much less likely to become deranged by tedium, which means your appendix is much less likely to swell up until it either kills or bankrupts you.

Apparently there used to be an aphorism in medicine that vestigial structures are particularly prone to disease, which sounds almost like a moral judgement on the feckless appendix. But in the developing world, the appendix is neither vestigial nor prone to disease. Only in the developed world is it one or both. So, to me, it emerges as a sort of tragic figure, a frontiersman out of time. I'm reminded of the 1996 Michael Bay action film *The Rock*, in which a group of US marines, who have fought bravely overseas to defend their country, go rogue and seize control of Alcatraz Island after returning from war to discover that our complacent modern society has no respect for their sacrifice. On reflection, *The Rock* is a better analogy for the appendix than either *The Wages of Fear* or my virginity.

When I used to have a mental picture of appendix-bursting, it was generally about twenty minutes into a promising first date. However, there are worse places. In 1960, a ship sailed from the Soviet Union to Antarctica to construct a new polar base at Schirmacher Oasis. The crew finished the job just as winter descended and the sea froze over. Weeks later, the base's doctor, a 27-year-old from Leningrad called Leonid Ivanovich Rogozov, fell ill. He realised he had acute appendicitis and there was no one else on the base who knew how to perform surgery. So, in an operation lasting nearly two hours, using only local anaesthetic, working mostly by touch because he couldn't see into his own chest cavity, he removed his own appendix. He survived the operation, and was back on duty within two weeks.

I imagine that Leonid Rogozov must have wondered why his appendix had to choose precisely that month of his life to make its rebellion. I imagine that from then on he had a deep faith in Sod's Law, or what Jerome K. Jerome called 'the natural cussedness of things in general'. I do too, and that's why I used to think it was inevitable that my appendix would disintegrate at the most inconvenient possible moment in the most expensive possible country. If this organ that I never wanted in the first place, this organ that had never done anything for me, this dead weight, this redundant curlicue, were precisely the organ that ended up having the gravest impact on my life, that would have a quality of the comic, the gratuitous, the senseless, the spiteful, that seems to me consonant with how the universe works a lot of the time. But now that I know the appendix isn't vestigial, that it's just a run-of-the-mill everyday organ like all the rest of them, I feel much more willing to acknowledge the run-of-the-mill everyday statistic that only about 7 per cent of people in the UK suffer from appendicitis in their lifetimes. Yes, the appendix is a relic and a liability – but only to the same degree as every other inch of the shambling, flaking, leaking, throbbing concatenation of relics and liabilities that I proudly call my mortal body.

EYE

ABI CURTIS

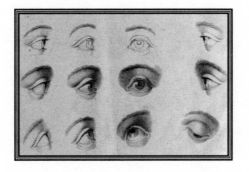

The room is dark. I see two lozenges of light on the far wall, one red and one green. My chin sits on a plastic rest. A voice tells me to look at the light. A shutter snaps down and I am left with an impression: a pattern of tree-roots radiating in a patch of brightness. I follow a torch with my gaze. In the almost-black, a man's silhouetted head drifts back and forth. I hear his breath and the hairs on my neck lift. In this strange intimacy, I see the shadows of the blood vessels belonging to my eyes. A series of lenses are clicked into place and I watch a chart of letters blur and come into focus. Which are real: the clear H, L, V and Z, or the dissolving versions that exist when the lenses are removed? I am twelve years old. I have started to become shortsighted. The blackboard at school is a dusty landscape and my mother has observed me narrowing my eyes like a cat. When I leave the optician's with new glasses I walk through a car park and am stunned by the smudges of grime on the windscreens. The world had been softened, and now it has come back in all its grubby imperfection.

My baby, when he was first born, could only make out

large shapes, movement and bright colour. His field of vision extended about twelve centimetres, a perfect focus for my face, my skin. As he developed he could see further and in more detail, and now he catches the trails of aeroplanes that I can barely see, or the ghost of a daytime moon. Our seeing creates a boundary. But seeing is not simply the practicality of looking and of navigating. Of our five senses, vision dominates and it is intimately bound up with our consciousness.

Astronomer Percival Lowell, in the late nineteenth century, was convinced of the existence of canals on Mars. Through his powerful telescope he saw evidence of a civilisation. The canals appeared cultivated, and with that came the thrilling possibility of a society. It turns out that the set-up of his telescope meant that what he thought was evidence of a community on Mars was actually a view of the complex and beautiful network of blood vessels at the back of his own eye. I have always found this rather poignant. Lowell insisted on his discovery up to his death and was not believed. The eyes give us our connection to the world around us, but they represent a strange loneliness too, there as they sit in our skulls, looking out from the singularity that is our brain: what we see is unique to us. Lowell was right in one sense: in some respects the eye is where society is created: in the connectivity with others that vision can allow. And the eyes remind me of my isolation, peering through mere screens at what goes on before me.

The ancient Greeks believed in the 'extramission' theory of sight: that it depended on the eyes firing out beams of light to illuminate, to 'touch', their surroundings. But why then, could we not see in the dark? This theory later gave way to the opposite notion of 'intromission', and in the *Book of Optics*, written a thousand years ago, the medieval Arab scholar Alhazen explained how the eye receives light. The eye draws the world to it.

I love to take photos of my travels on a big SLR camera, its twisting lens an extension of my flawed vision. So I also love how, in the early seventeenth century, Johannes Kepler explained how our eyes are like the camera obscura, uncanny, vitreous gadgetry for recording the world. An optometrist once showed me a photograph of the inside of my eye. The image was distinctly planetary: a haunting, pinkish globe, pulsing with web-like blood vessels, and perhaps it is no surprise that Kepler was also an astronomer, eager to understand the mysteries of celestial bodies. Lens, retina, pupil, stoma; fleshy, organic camera obscura, tethered by the threads of the optic nerve to the brain: the plush darkroom that develops its images in a split second. While the eye is stunningly complex, it incorporates an intriguing 'design' flaw: the blind spot caused by the structure of the optic nerve – the *punctum caecum*. We all have a point of blindness of which we are unaware.

When we make 'eye contact', the organ viewed from the outside is a gorgeous thing, the irises blue, green, brown, hazel, grey – those threads of colour like a gathering storm or tilled earth, all converging on the black hole of the pupil. I love to decorate my eyes with glittering shadow, mascara and kohl, framing the places in my skull from which I gaze. An eye that sees well is shaped like a globe but, being short-sighted, my eyes are flattened slightly, like a disc. If you are long-sighted, your eyes are shaped more like a torpedo or a lemon. This minor imperfection makes my mornings a fuzzy confusion, but modern science means I can live my life normally. I remember, aged 15, I was allowed to swap my heavy glasses for thin, plastic lenses that floated on my eyes. I was living in Hong Kong at the time, and as I left the optician's, with no part of my vision obscured, I twisted my head in wonder to the hot, mir-rored skyscrapers: the world was new again.

But I can't think about the eye without exploring blind-ness. I was deeply moved by the story of the theologian John Hull, who began to go blind in the 1980s. I sat in my office as the dusk lowered, listening to Hull's taped account, imagining him at his own desk, speaking softly into the recorder. His journey into blindness is also an exploration of grief and of a changing sense of conscious-ness. Hull wonders if the loss of faces in particular begins to loosen the connection to the self. His wife, Marilyn, reflects, 'I can't look into his eyes and be seen. There's … no beholding in that sense of being held in somebody's look … When you're very close to somebody that is a huge loss.' The blindness not only affects the blind, but the loved one who can no longer *be seen*. I imagine not being able to see my child, his growing, changing features. I imagine my husband not being seen by me, searching my face for recognition of his own. I am heartbroken by Hull's narrative of a dream in which he sees the face of one of his children who was born after he had gone blind – a face he will never see but one his unconscious has imagined for him.

Eventually Hull begins to accept his blindness as a 'dark, paradoxical gift'. He gives a beautiful reflection on the way in which rain illuminates the shapes of the outside world, and wonders, 'If only there could be something equivalent to rain falling inside; then the whole of a room would take on shape and dimension.' Rain could be a kind of seeing. There are, I realised, other forms of seeing and of knowing. Hull's depiction is of a paradigm shift where the intimacy of blindness alters one's consciousness.

I went to my local teaching hospital to meet a health-care assistant who works in ophthalmology and whom I shall call 'Greg'. Greg approached me in the busy waiting room and confidently shook my hand. I wondered for a moment if I had the right person. Greg lost his sight, due to a stroke, twenty years ago. As a result he now only has peripheral vision. As he described it, people's faces appear blurred as they do in police documentaries, when identities are obscured. I pondered this as I talked to him; that to him my face was a disc of blankness – I was disguised.

In this conversation, I was unseen. And yet, you would not know Greg was blind, and that's how he prefers it. He admitted to periods of anger and resentment in the first years of his blindness, a reaction that sounded a lot like grief. He refused to use the white cane issued to him, saying ruefully that such a thing in Merseyside, where he lived at the time, was an invitation to be mugged. Greg has a genuine understanding of his patients' physical and emotional experience. He has accepted his 'dark, paradoxical gift', but not without a period of mourning, of letting go. He told me about a patient he once encountered: a man in his thirties, blind since birth, who said he would never want his vision back. For Greg, who dreams of an advance in science that will return his sight, this was a strange revelation. As John Hull says, 'Sighted people live in a world that is a projection of their sighted bodies. It is not *the* world, it is a world.' What became clear to me was the complexity and delicacy of the eye: much can go wrong; those beautiful, root-like blood vessels are deeply vulnerable. They overgrow, like vines, bursting and leaking in their attempts to heal. The optic nerve, like a fleshy broadband fibre, communicates images to the visual cortex, that red-curtained movie theatre of the brain. If the nerve is damaged by a loss of blood supply, blindness is usually irretrievable.

If the optic nerve is not damaged, and cataracts are the cause of blindness, there are more options available.

Cataracts are cloudy patches which develop on the eye lens. The word 'cataract' has an oddly poetic etymology deriving from the Greek for 'portcullis' or gate, but the word has an alternative meaning: that of the waterfall, and the dynamic sense of 'down-rushing'. The 'caterickes' mentioned in *King Lear* are ostensibly part of the raging storm, but can also be interpreted as a pun on Lear's moral blindness: 'Blow wind & cracke your cheeks, rage, blow / You caterickes & Hiracanios.' This other watery sense of the word intrigues me. After all, cataracts are a portcullis for their bearer, a barrier obscuring the world from view; but for the outside observer they can look like the depths of a swirling waterfall.

Cataracts can be cured with a simple, seemingly miraculous, operation. Some of the staff at the hospital told me that it was witnessing this miracle that inspired them to work with eyes. But is the experience of having one's sight restored just as simple? The relationship between the eye and the brain is complex. In 1688 Irish scientist William Molyneux, whose wife was blind, wondered what would happen if someone born blind had their sight restored. How would they know forms and shapes that the brain had never had to interpret? How would they connect the sense of touch to the sense of vision, the two having developed separately? The known world is not necessarily the visual world. The neurologist Oliver Sacks recounts the case of 'Virgil', blind since childhood, who regains his sight after a cataract operation in middle age. Virgil's experience is not straightforward. The new shapes and lines before him do not resolve into structures, buildings or paths that he can navigate. There is no sense of perspective. He knows what steps are, but cannot walk up or down them. It is like trying to exist in the paradox of an Escher painting.

At the hospital, I considered the squashy vulnerability of the eye, how patients must overcome a squeamishness to allow it to be injected, touched, probed. During the Second World War, the ophthalmologist Sir Harold Ridley examined wounded pilots with shattered pieces of windscreens embedded in their eyes. He realised that, unlike glass, shards of acrylic were not rejected by the eye. This led to the discovery that clouded lenses could be replaced with plastic ones in what has become a simple medical procedure. The writer John Berger describes the removal of his cataracts as 'comparable with the removal of a kind of forgetfulness … a kind of visual renaissance'.[1] Experiences Berger did not know he had forgotten make a kind of return: 'the exact grey of the sky in a certain direction, the way a knuckle creases when a hand is relaxed, the slope of a green field on the far side of a house – such details reassume a forgotten significance'.[2] This kind of blindness is so easily cured, but is a miracle of the privileged with access to surgeons. Most of us take our sightedness for granted, and our sense of who we are comes almost casually, slyly, bound up with that. Blindness, especially when it befalls one later than early childhood, can feel at first like a bereavement and a strange, intimate trap: an oubliette down which one is forced to fall. For those to whom this has happened, their very consciousness must be remade.

Some of our contemporary seeing is virtual – and indeed virtual reality can offer us visions of the hyper-real, of a landscape that does not physically exist, and the eye will accept the illusion. I, like many others, glance at any number of screens on a daily basis, sometimes as the physical world scrolls by unseen. For my son, who is now two, the visual world is a continual surprise. Together we have renamed the everyday: a full moon is a 'wolf moon'; a crescent moon is a 'story book moon'. Through my toddler's perception I am seeing with fresh eyes. I think of Greg, seeing only the very edges of things, and remind myself to look up, from time to time, to see the charging clouds, the texture of leaves on the path, the delicate colours of the irises looking up at me, blue, dappled with green and grey, stormy with light.

Notes

1. (Berger, J. 2011:42)
2. (60)

Sources:

Berger, John (2011) *Cataract*, Notting Hill Editions

Francis, Gavin (2015) *Adventures in Human Being*, Wellcome Collection

Hull, John (2017) *Notes on Blindness*, Wellcome Collection

Middleton, Pete & Spinney, James (dir.) (2016) *Notes on Blindness,* Artificial Eye

Sacks, Oliver (1995) *An Anthropologist on Mars,* Pan MacMillan

Sheehan, William (2003) 'Venus Spokes: An Explanation at Last?' in *Sky & Telescope: The Essential Guide to Astronomy,* http://www.skyandtelescope.com/astronomy-news/venus-spokes-an-explanation-at-last/, accessed 12.12.2017

The Vision Eye Institute – 'The amazing World War II discovery that led to modern cataract surgery' https://visioneyeinstitute.com.au/eyematters/amazing-world-war-ii-discovery-led-modern-cataract-surgery/ accessed 12.12.2017

BLOOD

KAYO CHINGONYI

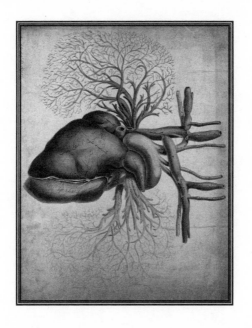

If I meet someone new, let's say at a dinner party, and that person asks about my family or my childhood, then I test how little I can get away with saying. This works exceptionally well if I'm sat next to an extroverted, chatty, person who, if I make the right sounds from time to time, will happily fill in the gaps. 'I'm told you're a poet; that must be difficult. I mean how does anybody make any money doing that? But I hear you have a book and you're doing really well so you must be doing something right?' It is fascinating to note how one-sided an exchange can be, how few words you can speak and still be in a conversation with someone. If I'm feeling particularly mischievous I make things up. The trick is to try and keep a straight face. If you can say something with a straight face even the most incredulous of sceptics will be taken in for a moment.

Sometimes I don't have to fabricate anything; someone will make an assumption about my life and I simply won't correct them. So when they ask where I grew up and I tell them, and then they imagine me living there with my parents, it is easier to stick to what they believe to be true than to explain myself further. Talking to someone for the first time is rarely the right moment to share the most painful facts about my life. But the things that come up when two people are trying to get to know each other, when both parties are attempting to appear breezy and casual, are very difficult for me to talk about. If I tell the truth, there is no way of making that truth *light*. If someone asks about my mum and dad, I can't say to them 'They both died when I was a kid.' Or rather, I can, but if I do, that immediately pushes the conversation in a certain direction. Parents are not *supposed* to die so early in their children's lives, and so sharing one of the most basic truths of my life means causing other people discomfort. 'How did they die?' is a natural follow-up question but, in this country at least – where restraint is an art form – nobody ever asks me that, though their facial expressions some-times do. Faced with a choice between letting someone I don't know believe I have lived a life I have not lived and watching their face change as they struggle to know what to say or do, I more often than not choose the lie.

It is that question, of how my parents died, that brings me to the subject of blood. If I'm in a forthcoming mood and someone asks me directly, 'How did your parents die?' I might say that my parents died from 'a blood con-dition', which takes care of the salient points but is merci-fully vague. What I rarely say is that both my parents died from bronchopneumonia because they had HIV. I never elaborate and say that they died because HIV impedes the function of white blood cells, the cells that help the body protect itself.[1] I don't go on to explain that in Zambia, where I was born, 'the prevalence of HIV among adults is thought to be somewhere between 12 per cent and 20 per cent depending on the region'.[2] I don't say that Zambia is a country where around half a million children have lost a parent or both parents to the virus.[3] The reason I don't, is shame. There is so much stigma, lots of which is racialised in deeply problematic ways, and such a dearth of understanding around HIV, that I don't feel able to say what is true without incurring judgement.

Perhaps my fear of judgement is something worth challenging. Perhaps people are more understanding than the wider culture would have us believe. But the risk involved in moving beyond that fear still feels significant. It has taken a long time to tell even those closest to me about the effect HIV has had on the course of my life. I knew my father had the virus and died as a result because my mum explained this to me when she felt I was old enough to understand. I did not realise when my mum started to get ill that it was the same virus; she never explained to me why she was losing so much weight or why she needed so many tests when I came with her to the hospital. As I understand it now, it's not something she accepted. She refused treatments that might have kept her alive. Did she do this out of shame? She is not here to answer that question and I can only guess at what she was feeling in what must have been an extraordinarily painful time for her. But if shame is part of the reason she refused help, then I have to knock shame on its head.

What my mum couldn't tell me as she faded from life she left to my aunt and uncle to explain. After taking time to let me adjust, to grieve, my aunt took me aside one day and asked if I had ever been curious about how my mum had died. In the way she phrased her question there was space for me to say 'no' but, as it happens, I said that I was curious; I did have questions. I am struck afresh, in writing this, by my aunt's capacity to defer a conversation until the right time. She looked at me then with the same calm expression as she had when she had previously told me a difficult truth, sitting with me before school not long after my thirteenth birthday, telling me that in the early hours of that morning my mum's lungs had collapsed and the medical staff had been unable to revive her.

After the second, more detailed, conversation with my aunt, I became obsessed with the notion that I had HIV too. It was possible to pass the virus from mother to child, through the blood, wasn't it? What if it had been going untreated all this time? I realise now that this fixation was my first attempt to come to terms with how I felt about the manner in which my parents died and, beyond that, how I felt about the fact that the only thing many people know about Zambia, if they know anything, is that it has one of the highest HIV infection rates in the world. I felt angry that my family wasn't an exception to that statistic, that it was not only my parents but their friends, neighbours and family members who had been affected. How could I hold that much shame in my body? I had to find out if I had the virus, too, and put the questions in my mind to rest.

I have, for as long as I can remember, been afraid of needles. When I say afraid I mean that sometimes my body spasms unexpectedly when images of needles pop, uninvited, into my head. When I say afraid I mean once when I was four or five years old, when I was taken to the doctor for booster injections, I took one look at the needle that was being primed for contact with my flesh and I ran: out of the room, out of the medical centre and a good way down the road before anyone was able to catch up with me. I was caught at the point when I was trying to decide if it was wise to cross the road on my own.

When I finally worked up the courage to be tested for HIV I was an undergraduate, studying for a degree in English literature. The university had been very conscientious about letting us know that it had a free, easily accessible, confidential, walk-in sexual health service. I took myself there early one morning and sat in the waiting room trying not to catch anyone's eye or mentally speculate as to why they were there. If I didn't do it to them, I reasoned, they wouldn't do it to me. In time my name was called and I was ushered into a room where a doctor asked me some questions. Was I sexually active? I was not. So what had brought me there? I explained. For the first time I was saying the words aloud to someone else. I was told it was possible I hadn't contracted the virus – there was no way of knowing whether my mum was HIV positive during pregnancy and even if she had been, mother-to-child transmission rates ('without medical intervention') range from 15 to 45 per cent.[4] They took my blood and sent it off to be tested.

As I walked home I thought about what it would mean to get the results. I knew the doctor was right and that I was unlikely to have HIV but what would I do if I did? Would I tell anybody? I walked the long way home and when I finally got back to my room I sat down against the wall – woodchip painted some variation of white, harbouring a nest of ants. I sat against the ant-infested wall and waited for my phone to ring. To pass the time I ran all the possible outcomes round my head in a fruitless loop. I didn't tell anyone what was going on because I didn't feel able to explain. So I sat in that room on my own conducting my unscientific study of time and its relative speed. After what felt like several hours but was, in all likelihood, about ninety minutes, somebody called to talk through my results.

I was not HIV positive.

Getting that confirmation gave me perspective. Taking the test was the first step in talking about how HIV has affected my life. It opened a space for dialogue that I had been shutting down. It is almost thirteen years since that day and it is only now I'm starting to see that what I have been calling 'a blood condition' is nothing to be ashamed of, that letting go of the shame I have around it lifts a heavy weight from my life. If, and only if, I can accept the things that have happened, then I can be truly present; then when someone asks me about my parents I can tell them that my parents met at university; that they fell in love; that they died when I was a kid; that it hurts a little every day; that even though it hurts I am still here and so long as I am I try not to let that pain be all that I can feel.

Notes

1. Terrence Higgins Trust, 'The Immune System and HIV'
2. UNAIDS, 'Zambia HIV and AIDS estimates (2015)'
3. Ibid.
4. World Health Organization, 'Mother-to-child transmission'

GALL BLADDER

MARK RAVENHILL

That first night in Warsaw two years ago I suffered a huge pressure beneath my breast bone that wouldn't go away. I shifted position in the bed, walked up and down the room, tried to breathe as deeply as I could. But still an invisible fist was pushing into my chest. I grunted and groaned, unable to sleep. It was, I guessed, severe indigestion.

The next day I was teaching a group of young Polish playwrights. The pain had passed but I'd had about forty-five minutes sleep. So I was a little slow in thought and speech but pleased to be past the pain and looking forward to a week of working with the rising talents of Polish theatre.

But that night in my hotel room the pain came just as sharply again. It returned, an unwelcome visitor, night after night all week. A deep constant punch, maybe softening for a few minutes, sometimes enough to doze, but always returning with the same merciless insistence. Inexplicably, but thankfully, the days were always pain-free. But as the week wore on I was teaching in a state of almost hallucinogenic sleep deprivation and had given up eating altogether, hoping that without food there could be no indigestion.

On my last morning in Warsaw, yellow eye balls stared back at me from the mirror in the bathroom. And I could detect the beginnings of a yellowing of the skin across my whole body. My urine was almost brown, my stools bordering on chalky white. I was jaundiced. A brief search on Google convinced me that I now needed to upgrade my self-diagnosis from indigestion to an advanced cancer. A driver came to pick me up at the Warsaw hotel. Did he notice my progressive yellowness, I wondered? I watched him closely but couldn't tell. As we drove the snow began to fall. Not a half-hearted sleet but great thick clumps of white until, as we reached the airport, it was a blizzard – stick your hand out to the full length of your arm and you wouldn't be able to see it any more.

Rushing to the check-in desk, I asked timidly (and thinking that I might now have to admit myself to a Warsaw hospital): 'Is there any chance that the plane will fly?' The woman from the Polish airline snorted, probably having heard tales of English airports closed at the first light dusting of snow. 'Of course plane will fly.'

Landing at Heathrow I asked a cab to take me straight to A&E.

'It's a gallstone,' the junior doctor told me.

'So not cancer?'

'Oh no, definitely not cancer. We'll send you up to the

ward now and in the morning they'll remove the gallstone from your pancreas. You should be out by lunchtime.'

'Good morning, I'm your surgeon,' said the surgeon the next day. 'I'll be seeing to the gallstone and while we're in there we'll take out the gall bladder.' He said, 'Once one stone has got out and into the rest of your body it's likely to happen again. Might as well make sure it doesn't.'

'Lose my gall bladder? Will I be able to lead a normal life?'

'Oh yes. The gall bladder's completely useless. If it's going to be a problem, best just to take it out. I'll see you later. You won't see me.'

I reached to google 'gall' and 'gall bladder' but found my smartphone's battery had died. Could it really be that the gall bladder is unnecessary? Wasn't gall once thought to be an essential human fluid? Wasn't it – I tried to remember a distant university lecture about Tillyard and the Eliza-bethan world view – wasn't it one of the humours? Yes it was. Blood, phlegm, gall and – um – something else were once thought to be the four fluids that moved about the body and whose equilibrium was essential for both bodily and mental health.

O the tragic fall of the gall bladder! Only a few centu-ries ago, responsible for pumping one of the four essential humours around the body. Now whipped out in a short session of keyhole surgery and – I'd guess – incinerated somewhere round the back of the hospital.

Preparing for this essay I arranged to meet Andrew Jenkinson, a surgeon at London's University College Hospital. He'd offered to let me in to his operating theatre to watch the removal of a gall bladder. I was relieved when he texted me early that morning to say that the gall bladder surgery had been delayed until another day. An emergency case had presented itself. A patient who'd had a gastric band fitted some months before had suffered a serious complication. The band had become twisted and she'd moved in the course of a year from being obese to dangerously underweight and had been rushed to the top of the operating list. So no operation that would be suitable for me to watch.

At the end of his working day, I met Jenkinson in the UCH's cafeteria. As he chewed on a nicotine gum – 'Want one?' he offered – he scribbled out a diagram for me, explaining the workings of the digestive system and the gall bladder's part in it. First he sketched the stomach – I was surprised to be reminded how high up it sat in the body, some way above where I would locate my belly – and then the liver (surprisingly big) and beneath it, like a little deflated balloon, the gall bladder.

The body produces gall (or bile, as we now call it) in order to break down fatty food in the stomach. The gall bladder doesn't produce the bile itself – it comes from the liver – but acts as a pump. Eat an extra-large cheesy pizza, say, and the body needs a sudden rush of bile to the stomach. So the gall bladder kicks in and pumps the bile in to break down the quattro formaggio. But the bile can crystallise inside the gall bladder, creating gall stones. These can cause discomfort if they remain inside the gall bladder but if they get out they can start to form blockages in the liver or, as in my case, the pancreas. Then it all gets very nasty.

So am I now less able to break down fats than I was when I had a gall bladder? 'There's some evidence that a very small number of patients suffer diarrhoea after gall bladder removal,' Jenkinson told me. 'Their bodies can't break down fat as efficiently. But it's very rare.'

'Then why have such an expendable body part?' I asked. I thought that evolution had ensured that we had efficient, almost utilitarian bodies. But, Jenkinson told me, human civilisation has moved much faster than the pace of evolution. We haven't – on a digestive level – caught up with the beginnings of human farming, tens of thousands of years ago. Our digestive system is still that of a hunter-gatherer.

The hunter-gatherer, Jenkinson explained, didn't eat the more or less constant stream of food that we do. Feast or famine was the general rule. Maybe once a week a bison might be speared. And then huge quantities of protein and fat would be consumed very quickly, with the pumping action of the gall bladder really coming into play. A fruit feast might come along a few days later, but then there'd be another sizeable gap before the body needed to break down huge amounts of fat and store it as energy.

So if medical technology advanced to the stage where we could push a button and delete the gall bladder, would Jenkinson recommend that everyone should have their gall bladder removed? 'If we could be sure that there'd be no complications,' he said, 'yes.'

Warming to his theme, Jenkinson pulled the piece of paper on which he'd sketched the human digestive system for me back to his side of the table. He marked a cross through the gall bladder and began to draw over his initial sketch of the stomach. 'In fact,' he said, 'we hardly need the stomach either. We've got stomachs that are far too large, that are meant to be full very rarely. But now that we're not hunter-gatherers the problem is that we can get access to food constantly. We fill the stomach far more often than we should.'

I shifted uncomfortably in my seat, aware that I'd achieved what I'd call a middle-aged spread, but what a doctor is more likely to label as borderline obesity. Jenkinson was wiry – probably about my age but with the body of a swimmer or cyclist and clearly someone who practices what he preaches. There aren't any excess calories going into that body, I thought ruefully, and promised myself to instantly begin a diet and exercise regime.

Jenkinson pushed the piece of paper back across the table to me. 'With our contemporary access to food we can eat smaller amounts regularly,' he says. 'We only need about 10 per cent of the stomach's capacity.' I looked down. He'd drawn a dotted line to create a thin tube of a stomach, cut free from the redundant 90 per cent. I looked across the table at Jenkinson. I detected an excitement in his eyes, and imagined his almost evangelical thrill at the possibility that the human being need no longer be stuck with a pre-agricultural body in a post-industrial age – that we can modify and remove until we have a body that's fit for the age we live in.

Clearly the technology is at a crude stage. Jenkinson had spent the better part of his day sorting out a gastric band that had led to terrible complications. But we probably aren't that far off the point where we can press delete and lose 90 per cent of our stomachs. Middle-aged spread or borderline obesity will no longer be an issue.

And I have to say, I haven't missed my gall bladder. If I'd been asked before it went, I'd have said that all of my body was an essential part of who I am. Well, maybe not the body fat that I'm constantly trying to lose – that's an unwelcome alien who's laid claim to my naturally slim body. But I still feel as though the hair on my head is an essential part of me – even though it was lost some twenty years ago to male pattern baldness. It's a strange and shifting thing – this sense I have that I am my body, of which some bits are essential and some expendable.

I still have my tonsils. I was just a little too young to

undergo the almost automatic removal that was considered essential for the generation before me. I was born in a nominally Church of England household so I held on to my foreskin. Although my appendix was removed when I was less than a year old, I can't say I've ever missed it. Bits of the body lost or not lost depending on culture, history, chance. My body, I realise, is not the stable thing I thought it was.

Since 1950, it has been mandatory for Australian explorers of the Antarctic to have their appendixes removed before their missions, to ensure that they don't suffer appendicitis far away from a surgeon. Similar prophylactic appendectomies have been fairly common but not mandatory practice for Russian, UK, French, Chilean and Argentinian explorers of the Antarctic.

In 2012, the *Canadian Journal of Surgery* published an article co-authored by a team of surgeons. With longer and longer space journeys planned – a colony on the moon and a manned mission to Mars no longer just the stuff of science fiction – should astronauts undergo prophylactic surgical procedures before leaving Earth's atmosphere? That is: should they have any unnecessary bits of them cut out just in case they cause medical complications far out in space? The report concludes as cautiously as you would expect from a group of Canadian surgeons:

As a result of the immense potential risk for loss of

mission and/or human life ... prophylactic surgical removal of a crew member's healthy appendix should be considered. This may also apply to a healthy gall-bladder ... the presence of gallstones clearly represents the greatest threat ... the ease and safety of surgical prophylaxis currently appears to outweigh the logistics of treating either acute appendicitis or cholecystitis during extended-duration space flight.

So on balance it's better for the astronaut to have their appendix and gall bladder removed – just in case. Could this, one day soon, be the standard advice for all of us?

I spoke recently to a female friend in the States who is considering a prophylactic double mastectomy. She's had no indications of breast cancer but thinks that after passing 50, and with a family history of the disease, it's better to be breastless. She has several female friends who've already had the same procedure. I nodded my support for her decision, supressing my instinct to say, But aren't your breasts an essential part of you, your womanhood, your beauty? Can you really just decide to let them go without the arrival of cancer? 'They're no use to me now,' she said with a sad smile. 'Might as well take them off for safety's sake.'

It's relatively easy to have the gall bladder or appendix removed – they've long since ceased to have much symbolic or cultural significance for us. But as medical technology becomes more sophisticated, we're going to face difficult questions. Which parts of ourselves are medically, psychologically, emotionally necessary? Am I my body? And how much of it do I want or need?

BOWEL

WILLIAM FIENNES

The pain began when I was eighteen: cramps like a torsion in the bowels, shock splashes of blood in the toilet bowl, the weakness after ten or twelve bouts of diarrhoea. I'd feel porous, ghostly, like cirrus, as if solid things could pass straight through me. When I was a child I thought illness was just an interval, at worst a few days in bed, my mother stirring glucose powder into fresh orange juice, using a kettle to fill the bedroom with steam, the world waiting outside until you were ready to step back into it. But this was a new region of experience and language: my abdomen inflating like a balloon as doctors pumped air in via sigmoidoscopes, plastic tubes threaded down my nose and throat into the stomach and ileum, litres of heavy barium milk betraying the sausagey coils of my intestines to x-rays, the 'sharp scratch' mantra of phlebotomists after fixing the tourniquet and pressing a latex finger to the vein, the companionship of drip-stands, the quick taste of metal before you went under; canulas and endoscopes; the splenic flexure and the Houston valve; ulcer, granuloma, Crohn's disease.

There are things we only think about when they go wrong: the fan belt, the combi boiler, the bowel. Before illness I must have imagined a gummy muddle behind my navel, but now gastroenterologists drew me a tube stretching 20 feet from mouth to anus, air and light at each end, an ingenious pipework that incorporated oesophagus, stomach, small intestine, ileum, colon and rectum, and contained 100 million nerve cells or neurons, more than in the spinal cord, as well as 95 per cent of the body's serotonin. I began to feel, specifically, the topography of the colon or 'large bowel' sitting across my abdomen – the ascending, sigmoid and descending colon, the bends at spleen and liver known as the splenic and hepatic flexures – which, when healthy, is a brilliant gourd absorbing 10 litres of liquid a day (water, saliva, gastric acids, biliary secretions, pancreatic juice) but which in my case had become the messy red bioscape of ulcers, inflammation and scar tissue I saw in photographs from colonoscopies, a tiny mobile eye with its miner's headlamp probing the dark, curving tunnels.

This vantage was strange enough, but soon I would see my bowel in a new light, not just at first hand but actually next to or beneath my hand, as surgeons cut a hole in my abdomen, above my right hip, and brought a loop of intestine out through it, slicing it open so the slurry of partly digested food known as chyme would stream straight out of my front into a bag. When I came round I could hear the bleeping of cardiac monitors along the ward and sometimes louder bleeps that seemed cause for concern but turned out to be nurses heating up ready-meals in the microwave. I wanted to see my stoma, the spout of intestine poking out above my hip; I wanted to *meet* it, this new presence in my life, already attributing to it an autonomous personality, as if it weren't really part of me at all. In the morning a nurse swept a curtain round my bed and drew the covers back, and I looked down at the small clear plastic bag smeared with blood and yellowish foam, and inside it a knob or bulb of wet, soft, pink-red tissue like the gums or tongue. The nurse saw that I was frightened and tried to reassure me; she said there were no nerves in it to register pain, that even if she inserted her finger into my side I wouldn't feel it. Dreamy on anaesthetic, I imagined her pushing first her finger and then her whole hand into the wound, up to the wrist, until she could have closed her fingers round my appendix or spleen and pulled them out of me, no blood to suggest anything unusual had occurred.

Of course you get used to almost anything, and although it seemed bizarre to be going home with a green plastic briefcase full of equipment – a selection of Coloplast, ConvaTec and Dansac bags, spare white plastic clips like barrettes a girl with long hair might use, Peri-Prep sterile wipes, special curved scissors for cutting holes in the card flanges – I soon fell into routines, kneeling at the toilet to empty the pouch; washing and drying; shoving all the dirty kit into the kind of odour-resistant plastic sack you'd use for nappies or dog messes, then fitting a fresh appliance over the stoma, pressing the flange to warm the adhesive, fixing a clip to close the opening. I'd never guessed how *interesting* it would be to keep tabs on these interior goings-on, to feel this bag hitched to my front like a sporran fill up with effluent that could be thick like porridge or watery like fruit juice or a minestrone bobbing with lettuce shreds and peas, to have such a window on my hidden basic processes, to see quite how much gas (or flatus, as ostomists learn to call it) came out of me during the night, so that in the morning my bag would be inflated like a zeppelin, straining at the glue, and in the bath would function as a buoyancy aid that drew my hips up to the surface of the water.

And then there was my fascination with the stoma itself, stoma from the Greek word meaning mouth, this pink teat above my right hip that sometimes relaxed and lengthened until it hung from my abdomen like a second penis. Once in the swing of changing the bags I learned that my stoma had moods, a mercurial personality – how sometimes it puckered into a bud like a nipple, tight against my skin, and sometimes loosened and elongated and became animated, feeling out the spaciousness beyond the body, so I'd think of an eel venturing out of its hole in a reef or the scene in *Alien* where the creature punches out of John Hurt's stomach to check out the mess room.

Depending on how recently I'd eaten and how relaxed I was, the stoma might be working, pushing out or dribbling chyme into the toilet pan, and sometimes I had a sense of appalled wonder that I should be able to watch my own inner functions like this, the muscular layers working in a wringing action to squeeze out liquid, peristaltic waves of muscle contractions moving the faecal stream through the bowels, and all of it going on beneath consciousness, by autonomic innervation, like the heartbeat ...

Fascination, yes. But also disgust. Sometimes the bag would peel off in the night and I'd wake to the smell and warm damp of my own excrement spreading across my stomach. Sometimes I stared at it in the mirror, this unnatural ornament, the pink bulb of gut stitched onto my side, the pouch of sewage I carried with me through the days. I imagined a myth in which, after some transgression or crime, gods forced a man to carry his shame around with him in a sack stitched onto his belly or side. I thought my stoma and its paraphernalia were the inverse of sensuality, a tag denying me access to erotic life; I couldn't imagine undressing in front of anybody, revealing the shitty baggage under my shirt; I couldn't even imagine holding someone, or dancing close against them, when they might feel, through clothes, the gummy knob above my hip, the plastic clip, the suspended, sludgy mass. I dreamed a woman I'd never met was kneeling in front of me. I wasn't wearing a bag, my stoma was clean and exposed – she leaned forward and kissed it, and I woke up almost breathless with the intimacy of that gesture. She might have been kissing my liver or the valves of my heart, places on the inside no one was ever meant to have access to, though there was a sort of tactile logic to that encounter of mucosal tissues, the bowel's vulnerable softness happy against the tongue's.

Sometimes I scanned the faces in crowds, wondering if there were others like me among them, other ostomists with their stashes of Peri-Prep wipes and carbon-filter flatus valves, and perhaps the newer two-part appliances with click-on Tupperware connections. Even though we couldn't recognise one another, we were members of a secret society, possessors of an organ most people would never encounter in their lives; we were the ones who knelt at toilets to empty ourselves and knew what it was like to defecate upwards from our navels during the night and how good it felt to stand in a shower with no bag glued to your skin, hot water streaming over the soft pink worm that lived in your side, a sensation of water sluicing down the *interior* of your body. And I thought of the moment the slit in my gut would be sewn up again and my tubing stuffed back into the cavity it had always belonged in, when my inner parts would be hidden once more and I'd no longer be this human Pompidou Centre that flaunted its pipes on the outside; I would be *whole*, as nature intended, my body restored to me. I would begin again.

I didn't know about prolapses, how some ostomists might experience their intestine spilling from the hole in their front like a sleeve pulled inside-out, so it was a shock to find one afternoon that my stoma was longer than usual – in fact, as I pulled the bag away from my body, I couldn't see the end of it at all, the mouth, just this long pink hose coiling up in the plastic. I dropped the bag and clutched the dangle of bowel against my abdomen, six or seven inches of it, thinking more and more of it might keep emerging, feeling I was coming undone, the stuffing emptying from the doll. I'd heard about early sorts of proto-stomas, soldiers hundreds of years ago who found their abdomens ripped open by musket balls and caught the tumble of guts in their hands. I'd seen paint-ings of St Erasmus, Roman persecutors using a windlass to wind his intestines out from a hole in his abdomen, so that man and contraption seemed attached by a taut umbilical rope. Later I dreamed my bowels were sliding out of me into a colander, slippery and hot, like tubes of cooked pasta. Soon paramedic Dawn in a green jumpsuit had me lying back on the bed. She stood over me and with her own gloved hands ushered the worm into my body. It slid back in until there wasn't even a protrud-ing teat, just a mouth flush to my skin like a whale's blowhole, and it seemed wrong somehow that nothing should have hurt, that this confusion of corporeal order and form should have taken place without any pain to indicate danger, an event for which evolution had never developed protocols.

Now this seems a long time ago, two years in my early twenties when I came to know my bowel as more than an abstraction. Days after the surgeon stitched up the little mouth and tucked it inside the crowded abdominal cavity, I stood in front of a mirror, peeled the dressing off and looked at my unpunctured torso – almost breathless, as if I'd been put back together, made whole again. I didn't guess that I would come to miss it sometimes, the weight of the pouch in my hand, the surprising sweet smell of the chyme, the handicraft pleasure of using curved ostomy scissors to cut bespoke holes in the bag flanges, the stoma itself like a rare pet of unpredictable modes and moods – puckered teat, loose dangle, worm making quizzical extensions into the world. I look down at the scar above my hip and think of it alive in its warm red nest beneath my skin.

KIDNEY

ANNIE FREUD

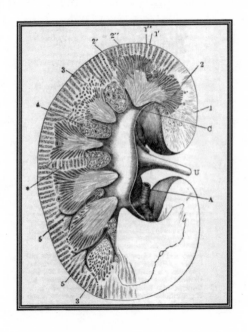

Apart from having once dissected a rat, a handful of literary references and a hazy childhood memory of a woman throwing a hairbrush at my father from her kidney-shaped dressing table, I hadn't given much thought to the reality of the kidney. So, my first action was to go to the butchers and buy some lamb's kidneys.

Once I'd got home, I was surprised not only by their softness but also by their slackness – the way that they behaved almost like a liquid mass, only just held together by their gossamer-fine membrane. How could something of such extraordinary complexity be so flabby? With almost no surface tension, only the very sharpest of blades would cut their flesh. I was also surprised by the apparent homogeneity of their interior substance.

And so, after a tasty steak-and-kidney pie with green peas and boiled potatoes, I arranged my two remaining kidneys on a plate in what I hoped was the kind of still life which invites the viewer to appreciate the anatomical facts *and* the aesthetic qualities of dead flesh – one kidney displayed whole, the other sliced into two halves. And then I started to paint.

I found myself using the richest maroon, the most delicate pink and the deepest crimson. I wondered what the purpose of my painting was. Finding no answer, I thought about the paintings of Chaim Soutine and Francis Bacon, the anatomical drawings of Da Vinci and Michelangelo and the meanings that have become attached to the depiction of butchered meat in art. I recalled seeing the kidney's fluent lines in the work of modernist architects and designers. Ernst Freud, the architect and my German Jewish grandfather, liked to include kidney-shaped fishponds in his garden designs. Although I still felt devoid of knowledge, I had handled them, sliced them, seasoned them, cooked them, eaten, stared at and painted them.

I spoke to several distinguished kidney specialists to find out more. I could not help noticing the admiration, passion even, with which they spoke of these organs, praising their diligence, complexity and versatility, using such expressions as 'fine tuning' and 'precise tailoring', and plying me with such wow-factor statistics as '25 per cent of the blood pumped by every heartbeat goes to the kidneys' and 'the whole 3 litres of our blood plasma is filtered by the kidneys forty-three times every twenty-four hours'.

My neighbour and friend Marcus Soldini, a GP with some twenty-five years practise behind him, asked me to imagine each kidney as comprising two distinct 'trees' – one the blood supply, the other the drainage system – whose outermost twigs are very densely interwoven.

'Visualise the blood,' he said, 'entering the arterial trunk, which divides and subdivides into ever smaller branches, each ending in a tiny knot of capillaries known as the glomerulus, of which there are about one million per kidney.' He gestured with his hands to show me how each glomerulus is embraced by the outermost twigs of the drainage system, forming a cup-shaped structure known as the Bowman's capsule, into which it fits snugly, like an acorn in an acorn cup. With peculiar emphasis, he said, 'This whole unit, known as the Nephron, whose dividing membrane is where the essential filtration process occurs – is one of the interfaces between the body and the external world.' I'm sure you'll know what I mean when I say that it was one of those 'thunderclap' moments, when the terminology with which some fragment of knowledge is conveyed feels like poetry.

And of course, it didn't stop there. 'Diabetes mellitus', the medical term given to the condition which, among other symptoms, causes the urine to have an abnormally high concentration of glucose, translated from the Latin literally means 'sweet fountain'. The amount of ultra-concentrated urine produced by the kidneys in states of extreme dehydration is called the *volume obligatoire*. Immediately I'm aware of an onrush of that slightly shameful disorder – so familiar to poets – in which the terminology of almost any endeavour, however remote, suddenly becomes invested with an irresistible glister and must, at any cost, be snaffled before anyone else gets their hands on it.

I was likewise enthralled by Marcus's description of the urine's journey back down through the drainage channels to the renal pelvis, from which a slender muscular tube known as the ureter slowly milks it into the bladder by way of peristaltic waves.

My instinctive response to the thought of writing about the kidney was to do with culinary pleasure; I remember a dark and stormy afternoon shortly before my first Christmas in Dorset. Wet through, ravaged with hunger and laden with bags, I saw the welcoming light of a small pub at the T-junction in the centre of Bridport. Kidneys on Toast was on the menu, priced at £4.95. Remembering the dish I ate that day with such relish, I thought of Elizabeth David's recipe:

Rognons Flambés
1 pig's kidney per person
Salt
Juniper berries
Ground black pepper
Dijon mustard
Cream
Brandy
Butter

Skin the kidneys and cut them in half. Put them in warm salted water for half an hour, cut them cross-ways into small pieces and season them with ground black pepper and a little salt.

Heat a little butter in a shallow pan and sauté the kidneys fairly rapidly. Turn them over so that they do not frizzle. After 5 minutes add 3 or 4 crushed juniper berries, pour over a small glass of brandy and set light to it. Shake the pan so that the flames spread. When the flames have gone out stir in 2 teaspoons of mustard mixed with 4 tablespoons of thick cream. Serve at once.

I found myself further transported, this time to the kitchen of Leopold Bloom in Joyce's *Ulysses*, who 'liked thick giblet soup, nutty gizzards, a stuffed roasted heart, liverslices fried with crust crumbs, fried hencod's roes ... grilled mutton kidneys which gave to his palate a fine tang of faintly-scented urine'. I can smell them burning in the frying pan and see Bloom himself, 'chewing with discern-ment the toothsome pliant meat'.

Rereading those words and enjoying their subversive profanation of kosher ritual, I am aware of a current vogue for offal on the menus of a certain kind of smart new res-taurant, for which the chef and writer Fergus Henderson has coined the phrase 'nose to tail eating'.

Singular tastes, lone experiences, private lusts ...

Although its etymology can't be traced exactly, the word 'kidney' originates from the fourteenth-century 'kidnere', which turns out – rather tellingly – to be a compound of two Old English words, 'cwid' meaning womb and 'ey' meaning egg. Thus: womb-egg! In comparison to the heart, the stomach or even the liver, all of which occupy much more centre-stage positions than the kidneys and have a myriad metaphorical uses, the kidneys, whether human or animal, appear much less often in history and literature and seem to have some quite singular associations.

In the Middle Ages you might have heard the expression 'a man of my (or his) kidney'. Back then the temperament of a man was believed to be governed by his humours, and the kidney thought to be the seat of the affections. Thus 'a man of my kidney' meant a person whose temperament and disposition were the same as those of the speaker. As spoken by Falstaff in Shakespeare's *Merry Wives of Windsor* and by the speaker in T. S. Eliot's poem 'A Cooking Egg', its use also implies the kind of buffoonery associated with setting too much store by one's status.

In Anne Desclos' *Histoire d'O*, the notorious sadomasochistic novel of the 1950s, the frequently repeated 'les reins' which, although literally means the kidneys, is used here with chilling ambiguity to designate 'the loins' as a euphemism for the female sexual organs. The suggestion is that which is innermost and therefore most vulnerable.

We find 'les reins' in poems by Baudelaire and Rimbaud in which they almost appear to be vying with one another to see which one most excels in the scabrous depiction of (female) carnal realities. The refrain of Serge Gains-bourg's duet with Jane Birkin, 'Je t'aime … moi non plus' is 'Je vais et je viens entre tes reins.'

A quite startling number of kidney references are also to be found in pop song lyrics: Al Jarreau, T-Bone Walker, Paul Weller, Marianne Faithful, Bjork, Mark E. Smith, the Red Hot Chili Peppers, Jay-Z and Eminem, among many others, have thought the kidneys worthy of inclusion. Special mention goes to Frank Zappa for this from the magnificent 'Pygmy Twylyte':

Doo-doo room
Reek replete
Crystal eye, crystal eye
Got a crystal kidney & he's afraid to die
In the pygmy twylyte.

That the kidneys receive more than thirty mentions in the Old Testament is perhaps not so remarkable until juxtaposed with the fact that the brain is not mentioned at all. According to Webster's *Concordance with the Bible*, the kidneys owe their importance partly to the fact that the fat that surrounds them was considered to be of great purity. So much so that it became a proverbial term for surpassing excellence, was regarded as best reserved for God in the sacrificial burning of animals and came to be deemed sacred.

The position of the kidneys makes them particularly inaccessible; they are the last organs to be reached when a butcher cuts up an animal. They assume the symbolic identity of the most hidden part of a man, and in the Book of Job to 'cleave the reins asunder' is to effect the total destruction of the individual. The hidden location coupled with sacrificial use caused the kidneys to be thought of as the seat of the innermost moral and emotional impulses. Thus the kidneys, or reins, 'instruct', are 'pricked', cause torment and rejoicing and come to symbolise man's conscience. To 'know' or 'try the reins' is an essential power of God, denoting his complete knowledge of every human being.

I knew that eventually I would have to include something about kidney disease. I am grateful to Hugo Williams for giving me permission to include one of his poignant and courageous poems in which he addresses the painful reality of his diabetes:

Diality
The shock of remembering,
having forgotten for a second,
that this isn't a cure,
but a kind of false health,

like drug addiction.
It performs the trick
of taking off the water
which builds up in your system,
bloating your body,
raising your blood pressure.
It sieves you clean of muck
for a day or two,
by means of a transparent tube
full of pinkish sand
hanging next to your machine.
Your kidneys like the idea
of not having to work any more
and gradually shut down,
leaving you dependent.
Then you stop peeing.
Dialysis is bad for you.
You feel sick
most of the time, until the end.
The shock of remembering,
having forgotten for a second.

Eventually I realised that my best reason for choosing
the kidneys has to be my husband, Dave. A few years ago
a malignant growth on one of his kidneys was successfully
treated with keyhole surgery. Luckily the growth was not
close to any other organs and was quite contained. It was

a very difficult year and I remember our relief in being shown the before and after images on the consultant's computer screen and his glee at the successful outcome. So I have a particular affection for my husband's kidneys and, since having devoted an unusual amount of time to thinking about these wonderful little organs, for kidneys in general.

BRAIN

PHILIP KERR

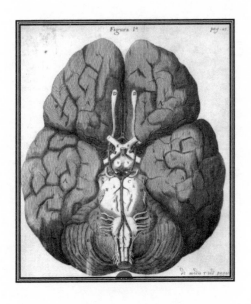

Breaking Bad is an American television drama about Walter White, an impecunious, middle-aged high school chemistry teacher diagnosed with lung cancer who, together with his former student, Jesse Pinkman, turns to a life of crime producing crystallised methamphetamine and selling it for large sums of money to secure his family's financial future before he dies. As Vince Gilligan, the show's creator, succinctly described the five-series story arc, 'You take Mr Chips and turn him into Scarface.'

Lobotomy was once the Scarface of medical procedures. It involves breaking the human brain, cutting or scraping away a lobe with the aim of treating the symptoms of a mental disorder, sometimes at the expense of personality and intellect.

My own aim with this short essay is to describe the procedure to you in a way that takes Scarface and turns him into Mr Chips. I hope to persuade you that this once notorious medical procedure is now a respectable one that brings hope to a great many people who suffer from temporal lobe epilepsy, and to restore the word 'lobotomised' to a position where it is no longer used as a pejorative term for someone who isn't very bright or has been rendered a vegetable by neurosurgical intervention.

But, as Walter White might say, 'We're getting ahead of ourselves here, Jesse.' We need to start with Scarface.

The first lobotomy, also known as a lobectomy, was undertaken in 1935 under the direction of the Portuguese neurologist António Egas Moniz. Many might wonder how a practice whereby you jam an ice pick behind someone's eye and break off bits of their brain like the parson's nose could ever become popular; but the use of the procedure increased dramatically in the early 1940s, and by 1951 almost 20,000 lobotomies had been performed in the United States alone. Egas Moniz was even awarded a Nobel Prize for Medicine in 1949 for discovering 'the therapeutic value of lobotomy in certain psychoses'. The procedure was always controversial, however, and not without its casualties. Following the introduction of antipsychotic medications in the mid-1950s, lobotomies were quickly and almost completely abandoned. So it's with these crude early lobotomies that the first part of my piece is concerned.

I'm sure you will be familiar with John F. Kennedy and the damage inflicted on *his* brain by Lee Harvey Oswald. But you may not know that JFK's younger sister, Rosemary, underwent one of the very first lobotomies in 1941 at the age of just 23. Perhaps she wasn't the brightest student at school, but her diaries reveal a thoughtful, observant young woman attempting to make her way in a large family run by one of the most ambitious and ruthless patriarchs of modern times, Joe Kennedy. Rosemary was assertive and rebellious, and doctors persuaded her father that a new and still experimental procedure would curb his wilful daughter's mood swings and unpredictable behaviour. He did not consult his wife, however, and it seems unlikely she would have given her consent. The account of the operation is horrific, to say the least.

They gave Rosemary only a mild tranquilliser. Dr James Watts made a surgical incision in the brain through the skull. The instrument they used to break off a piece of her brain looked like a butter knife. As Dr Watts cut, another doctor, Walter Freeman, put questions to Rosemary. He asked her to recite the Lord's Prayer and the two doctors made a guesstimate on how far to cut based not on an ECG (no such thing then existed) but, incredibly, on how she responded. Which is a little like Isaac Newton investigating the back of his own eye with a bare bodkin: foolhardy to say the least. When Rosemary's recitation became incoherent, they stopped. After the lobotomy it quickly became apparent that the operation had been nothing short of catastrophic. Rosemary's mental capacity diminished to that of a two-year-old child. She was immediately institutionalised and for the rest of her life she could neither speak nor walk, and was incontinent. Joe Kennedy never saw his daughter again. And it was another two decades before any of her siblings learned the truth about their sister's disappearance.

Like a lot of people, my own awareness of lobotomy came through literature and film. Tennessee Williams' older sister, also called Rose, underwent a lobotomy that left her incapacitated for life. The great playwright criticised the procedure in his play *Suddenly Last Summer*, and the way it was used in attempts to make homosexuals 'morally sane'. But it was probably Ken Kesey's 1962 novel *One Flew Over the Cuckoo's Nest* – later made into a 1975 film starring Jack Nicholson – that expanded the operation's notoriety by a factor of ten. In the story, the swashbuckling, rebellious and charismatic hero, Randle P. McMurphy, is given a lobotomy after he attacks the tyrannical head nurse in an Oregon state mental hospital. Kesey's narrator, Chief Bromden, describes the tragic result: 'The swelling had gone down enough in the eyes that they were open; they stared into the full light of the moon, open and undreaming, glazed from being open so long without blinking until they were like smudged fuses in a fuse box.' Another patient says of McMurphy: 'There's nothing in the face. Just like one of those store dummies ...' I still vividly remember the moment in the movie when the Chief lifts McMurphy's supine body tenderly, looks into his friend's vacant face and realises with horror that while the lights are on, no one is home. It's one of the most shocking moments in modern cinema. Almost as disturbing was the scene in the 1968 classic sci-fi movie *Planet of the Apes*, when the astronaut Taylor, played by Charlton Heston, discovers that future ape scientists have lobotomised his fellow crew member.

The fact sheet on lobotomies reads like this: doctors Freeman and Moniz – the early pioneers of what Freeman himself described as 'surgically induced childhood' – used the procedure as an attempt to cure things like schizophrenia, chronic headaches, migraines, post-natal depression, manic depression and mild behavioural disorders. Dr Freeman once performed a staggering twenty-five lobotomies in a single day, and it's hardly a surprise that many people did not fare well after the procedure. One boy, Howard Dully, received a lobotomy because his mother didn't like him. Several thousand US soldiers returning from the Second World War and suffering from post-traumatic stress disorder were lobotomised. Unless the procedure killed someone, the operation's practitioners considered all of the permanent brain damage and reduction to the status of a breathing vegetable to be nothing more than collateral damage of what they regarded as an otherwise effective treatment.

At this stage you might be forgiven for thinking that I've given myself a hopeless task, that it would be more or less impossible to turn Scarface into Mr Chips and claim that lobotomy is once again a respectable medical procedure. But it is.

What has changed is this: back in the 1940s and 1950s surgeons were poking around in the brain using their ice picks and butter knives with little or no idea of what they were doing, a bit like Christopher Columbus setting sail from Spain without a map and no idea of where he was really going or what he expected to find when he got there. Essentially it's the map-making that has changed. Thanks to X-rays, ECTs, MRIs, PET scans, SPECT scans, EEGs and DBS (deep brain stimulation), surgeons now have a much clearer idea of the brain's previously hermetic topography – what happens where and why. It's now possible to say, with certainty, what part of which lobe of the brain handles eyesight, smell, speech or movement. For example, the part of my brain that handles my delivery of a lecture, for better or worse, is Broca's area, a region in the frontal lobe of my brain, while the part of my brain that will tell me to shut up when I run out of time is the lateral intraparietal area, which is toward the back of my brain.

The great mysteries of this, the most wonderful human organ of all, are now finally being understood in a way that early cartographers of the brain and head like Franz Joseph Gall and Cesare Lombroso could only ever have dreamed of. Now, if I could only work out which part of my brain was mad enough to think I might be equal to the task of writing an essay on the subject of neurosurgery then, perhaps, we would know everything. We are now equipped with a *mappa mundi* of the human head that we might call an electronic *mappa cerebrum*. It's as if surgeons have the very latest satellite navigation system for any journey undertaken inside your skull. Whether it's a simple craniotomy, or drilling a burr hole to alleviate a subdural haematoma, all neurosurgical procedures can now be undertaken with much more confidence as to the treatment and outcome.

For reasons of sensitivity probably relating to its existing reputation, as earlier detailed, neurosurgeons now describe lobotomy as an anterior temporal lobectomy, or ATL. It involves the complete removal of the anterior portion of the temporal lobe of the brain and is now the standard treatment option for those patients with medically intractable medial TLE, or temporal lobe epilepsy, which is to say those in whom anticonvulsant drugs do not control epileptic seizures. The procedure is still risky and extremely expensive, but the range of seizure-free outcomes for these patients is reported to be between 80 and 90 per cent.

Breaking brain is no longer done with a butter knife or an ice pick. I can tell you first-hand because, a few weeks ago, your intrepid reporter went to see an ATL carried out at the National Hospital for Neurology and Neurosurgery in London's Queen Square. Gone are the days when twenty-five lobotomies could be carried out in one day. The single ATL procedure that I saw lasted almost eight hours, with as many as nine medical personnel involved, including three neurosurgeons. Prior to the surgery itself my attention was drawn to a tiny, almost invisible lesion or scar on an X-ray of the patient's head, which would probably have escaped the notice of a Swiss watchmaker but not our neurosurgeon, Dr McEvoy, who explained that the scarring to which the patient's seizures were certainly referable had probably occurred as the result of some high fever contracted as a child. In an atmosphere best described as the opposite of febrile the ATL procedure commenced.

These days it takes almost two hours of work with an electric scalpel, or Bovie, to remove a flap of muscle and clean away the surface of the skull prior to craniotomy. The craniotomy is then carried out with a high-speed drill which reminded me – both in sound and smell – of what happens when you have a filling at the dentist. A section of skull about the size of a matchbook is removed and stored safely so that it might be grafted back on later.

Under the skull a membrane known as the dura encloses the brain like the skin on a haggis; once cut it reveals a glistening grey brain covered in a cobweb of blood vessels that most resembles one of those delightful eggs in the movie *Alien*. Using a giant neurosurgery microscope that affords perfect visibility in deep and inaccessible cavities like the human skull, the delicate business of breaking brain can now be undertaken.

While he worked, the surgeon told me that after the lobectomy the brain would quickly rewire itself, and that the patient already regarded the short-term inconvenience of his brain having to establish new synapses as a price worth paying after a lifetime blighted by epileptic seizures. In my own lay opinion the whole highly technical and immensely delicate scientific procedure seemed like a great advance on guesstimates and ice picks, not to mention the Lord's Prayer. The patient, you will be glad to hear, is now recovering well and in full remission of the temporal lobe epilepsy that occasioned his neurosurgery.

And if Jesse Pinkman, Walter White's young partner in crime, had been there to witness this astonishing operation instead of me, I like to think he would have high-fived the neurosurgeon, and said, 'Yo, bitch. Science.'

LUNGS

DALJIT NAGRA

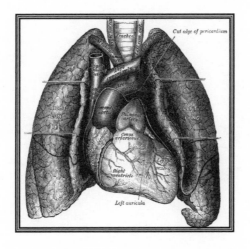

As one of the five million Britons who suffer from asthma, I've had a necessary interest in my lungs and an acute awareness of breathing from an early age. I remember the terror of early childhood asthma attacks and was recently shocked to learn that, even now, the condition kills three people in this country every day.

The recorded history of asthma goes back a long way; it is first mentioned as a medical condition by Hippocrates, over 2,500 years ago. Despite this, advances in medicines for asthma have been incredibly slow. It is only since the 1960s that we've seen the significant medical developments which have alleviated asthmatics' suffering. I've been treated with asthma since the early 1970s, and in that time I've lived through three generations of asthma medicines.

When I was a child I used to endure the indignity of milky-coloured bullet-shaped suppositories on a daily basis. Luckily, by the mid-1970s the spinhaler replaced this medication. The spinhaler was a groovy circular machine into which you cracked open a capsule, which the patient then sucked in from a mouthpiece, making a great wheezing sound. This became my party piece, a way of impressing my friends. Then along came an even more joyous way to relieve asthma: Ventolin. Just one puff from this inhaler and the adrenaline opened up my airways whilst also leaving me slightly high. I was a Ventolin addict through most of my teens. I'd go to my bedroom, sit back and look at the sky whilst leisurely puffing away at the inhaler. If I took enough puffs I'd see stars before my eyes. Inhalers were life-changing for asthmatics, giving us the comfort of knowing that our ever-constricting airways could be reopened in a puff and our lungs able to fill with air once again.

Before I was old enough to administer these Western medicines myself, however, I rarely used them; my Punjabi parents had no understanding of my prescriptions, as their lack of English meant that they were unable to communicate with doctors or read instructions. Besides, they were keener to find a traditional Eastern remedy for my asthma. My earliest memories of asthma, when I was five or six, involve being dragged by my parents to countless faith healers to help rid me of this dishonourable condition. I come from macho farming stock and, to compound matters, my father was a champion wrestler. And here I was, a wheezy wreck.

As I'm from a Sikh background the faith healers were always Sikh men who hid behind a uniform of saffron robes, a long beard and a turban. These men plied their craft from settings as diverse as concert halls, lofts and riverbanks, and their advice was just as varied. One prescribed a pot of black chicken broth twice daily for a month. Another whom we visited near the Golden Temple at Amritsar dunked my head in the holy river. One sternly informed my parents that I was carrying a curse from a previous life and decreed that I must atone by reciting prayers on an hourly basis. Perhaps the one I remember the most vividly is the man who issued my parents with two cotton threads to put through my ears, which had to be pierced especially. I was about five years of age at the time and was made to wear black cotton earrings. The sacred power of these rings sadly didn't last very long, as I had an asthma attack within a few days of first wearing them. I still have a small lump on each ear to remind me of this attempt at a cure.

None of the potions and spells of these faith healers worked and they left me unimpressed by the mystical world. As I grew up I became increasingly removed from the traditions and beliefs of my parents, and I now wonder how much of this was due to these disturbing childhood memories. Instead, I became a believer in the incantatory power of poetry. Whilst researching lungs at Guy's Hospital I came across a beautiful life-sized bronze statue of one of my favourite poets, John Keats. Keats trained as a physician at Guy's in London before dying in

1821, aged just 25, from tuberculosis, a deadly infection of the lungs.

I went to Guy's Hospital to look at some lungs with two pathologists, Ian Proctor and Elaine Borg, having already expanded my lung knowledge with Professor Douglas Robinson, consultant in respiratory medicine and allergy at University College London. Professor Robinson explained to me that the lungs are controlled by the rib cage expanding and the diaphragm moving down, forcing air into the vacuum created in the lungs which absorb the oxygen and expel the carbon dioxide. There is something quiet about the lung in this process; we never sense them at work. I love the idea of lungs having this almost ethereal property, of being here in an ever-so delicate and modest manner. As Ian Proctor explained to me how exception-ally light they are, I watched his eyes go upwards as if he were imagining a lung flying free of the body.

After our meeting, I learnt that 'lung' is an Old English word probably derived from German meaning light, as in not heavy; hence the lungs are known as 'the light organ'. In spite of this springy lightness, they are surprisingly strong, and would soon blunt a knife if sliced through, because each of the several million air sacs are encased and kept open by cartilage. If a lung and all of its air sacs were opened and spread out they would cover a tennis court. Yet thankfully we're not weighed down by these enor-mous organs, as their combined total weight, when fully functioning and circulating with blood, is about 800 grams, the approximate weight of a loaf of bread. When inflated, each lung has the volume of a football and fills a cavity from the base of the neck almost down to the waist. Lung capacity can actually be increased to improve the efficiency of our body, ensuring that oxygen is delivered faster to the blood and enabling our organs and muscles to work more effectively. For sports men and women, and singers and wind musicians, an increased lung capacity is essential for top performance. Apparently swimmers and bagpipers top the lung capacity league. Whilst the average human lung capacity is six litres, Olympic swimmer Michael Phelps' is double that, at twelve litres; however he has a long way to go and many oceans to swim before he catches up with the blue whale, whose lung capacity is 5,000 litres.

London's Gordon Museum of Pathology, which houses a great collection for medical students to refer to, is itself a huge air-filled cavity lined with walkways and shelf upon shelf of organs and body parts, each suspended quietly in its own sealed and numbered jar. There I saw several specimens of lungs. These were strikingly varied in colour. I was told that the lightest, cleanest ones belonged to country dwellers. Those of city people were speckled with black; the lungs absorb the environment in which they breathe. Of course, the lungs of smokers or those affected by cancer and tuberculosis were different again, appearing very dark and in some cases eaten away.

Looking at the diseased and infected lungs made me think back to Keats and his tragically early death from tuberculosis. He wrote his 'Odes', some of the most beautiful and enduring poetry ever created, over a few months in 1819 whilst at the house in Hampstead, which is now the Keats House Museum. Keats walked frequently on Hampstead Heath at this time, taking in its clean air and looking down on the pollution in the city below. To this day, the Heath is known as the Lung of London; we can visualise it absorbing all our excessive carbon dioxide and releasing oxygen for its millions of Londoners. I wonder how many other great cities have such a sprawling lung, and I think immediately of Manhattan; from above, Central Park's vast greenness dominates the island, located as if to ensure it is in the reach of all New Yorkers and that its oxygen can extend as far as possible into the complex grid of its streets.

As a poet, I'm inclined to think of the poem as a temporary breathing system that rewards the reader with its own riches. To me, absorbing a poem through reading is an exchange system which helps us remove the toils of the day and replenish us with beauty. More mechanically, the breath of poetry impacts on our lungs through its rhythms. I think of a line of verse, or a unit of sense being recited in a single breath, then the reader taking a pause for an inhalation of breath before the next line of verse is recited. I can usually feel each breath even when reading the poem silently in my own head. If I'm ever feeling unwell, I find it hard to edit one of my poems, because the lively breathing required can be quite exhausting. Again, I think back to Keats and the fact that even when he was ill with tuberculosis, he still found the mental energy to hurl himself into great lungfuls of lively verse which contains highly energetic, syntactically complex arguments.

When writing, I'm always keen to alter the regulated breathing of the reader so they feel they have lived in my poems. I want the reader to acquire my breath rhythm, so that for the duration of the poem my breathing becomes their breathing and they have been taken out of themselves, taken on a breath journey. The breathing system of one poet differs from that of another. The breathing experience will be a challenge for the reader when reading a section of, say, Milton's *Paradise Lost*, with its delayed verbs and its powerfully excessive system of minor and major clauses charging through its long sentences. In contrast, a few lines of Carol Ann Duffy's poems can seem a breath-doddle because of all their one-word sentences and frequently short clauses.

It has been argued that even the most familiar English line, the iambic pentameter, is not really a pentameter. Instead it's broken in two, with two strong beats either side of a weaker middle bit where a breath is taken. For example, consider the opening line of a Shakespearian sonnet: 'Shall I compare thee to a summer's day'. This could be read in a continuum with five regular beats or with a mid-pause: SHALL I COMPARE THEE to a SUMMER'S DAY, the weak pause between 'thee' and 'to' allowing for a slight intake of breath so we give life to the final two beats and allow for breath variation.

So much of the emotion, the mood of lyric poetry exists in the breath of the verse. The Black Mountain poet Charles Olson wrote in the 1950s that part of the power of a poem lies in the poet's control of the 'pressure of the breath', and I like the idea of 'pressure' being exerted so the poem becomes a song. The poem may not have drums and strings accompanying it, but it has its own rich patterned music of words which, if controlled by the poet, creates a breath-pressure, and when the reader feels this release and inhalation of breath they enter the music of the poem.

So the poem, with my newly acquired knowledge, is a physical event. A poem alters the reader's experience of their rib cage and their diaphragm. The poem can help create an air thrill in the thorax which can make the lungs feel calmed or rushed. I wonder if this is one of the many things the great Polish poet Zbigniew Herbert had in mind when he said that poetry should be a struggle for breath. Herbert, witnessing twentieth-century upheavals in Poland, saw the act of writing a poem as an attempt at survival in the face of social and political brutality. The completed poem, the enduring poem, a breath victory! So that, finally, over the oxygen and carbon dioxide exchange system of breathing rides the despair, the fortitude, the joy of the song, of the poem.

EAR

PATRICK MCGUINNESS

As a schoolchild, I was in equal measure spooked and puzzled by my encounter with literature's most famous ear: the scene in *Hamlet* where his father's ghost appears and tells him how he died. The ghost informs Hamlet that the official story – that a serpent bit him as he slept – is a lie, and in fact Claudius poisoned him by way of his ear:

> … thy uncle stole
> With juice of cursed hebenon in a vial,
> And in the porches of my ears did pour
> The leperous distilment, whose effect
> Holds such an enmity with blood of man
> That swift as quicksilver it courses through
> The natural gates and alleys of the body

Sitting side by side in classrooms, children get to see a lot of ears. The more observant ones might notice that ears are all different, but also basically the same: that folding, creasing, puckering skin over tiny bones, like rucked up sheets or crumpled tarpaulin. Sometimes, against a window or close to a light, you'd actually see *through* another child's ear, the skin seeming thin as ricepaper, translucent and veined as the sun streamed through and lit it.

My own ears, like those of most children, were repositories of earwax and the scenes of occasional infections. The ear is both prosaic, grotty and banal, and hyper-sophisticated, beautiful and complex. It was my ears that my schoolmates flicked and my teachers folded back to check if I'd cleaned behind them after sport. But the same ears let me hear Jacques Brel's 'Ne me quitte pas' at the age of 12, and dredged up tears from somewhere deep inside me.

Beyond the eardrum most of us know little about the ear. Maybe that's because the eardrum is where the cotton bud stops, or should stop. I once pushed mine too far in, in search of an elusive crust of earwax, and tore my eardrum. I screamed because I thought I'd drilled into my brain and it would leak out. It wasn't just the pain, it was the sense of having broken the barrier between the inside and the outside of my body.

Cleaning my ears gave me huge pleasure – in those days, we made our own fun. And although we're now advised against it, I suspect I'm not alone in still enjoying the tentative push and twirl as the cotton bud goes in, probing, scooping, finding the little encrusted crannies, turning around inside the earhole like a spoon finding the angle of the jar to scrape out the last of the jam. And the triumphant return, the earbud laden with treasure; or the disappointment if it comes back as clean as it went in. Hotels know this, which is why their welcome packs often contain a bizarre assemblage of items: shower cap, sewing kit, shoe-shine and earbuds. I like to imagine the sort of holiday I'd be having if I ended up using them all.

Before his revelation about the poison, the ghost of Hamlet's father says that 'the whole ear of Denmark ... is Rankly abused' – the entire country has been taken in by the lie about his death. Abusing a country's ear is perhaps an early version of fake news, that new name for an age-old phenomenon. But what makes the ear so powerfully symbolic for Shakespeare is the way it connects the inner world to the outer world not just biologically, anatomically, but cognitively and spiritually. The ear is an entrance, a portal, a porch not just into our bodies, but into our brains, where everything we call ourselves takes place. The ear is always open. Because it has no bodily off-switch, *we* have no off-switch either. Even sleep, the closest we have to an off-switch before the greatest off-switch of them all – death – is at the mercy of our ears.

What is the process by which we hear? Perhaps it is best described as the story of sound, or the journey a sound makes, because stories are journeys and, like the ear's three connecting spaces – outer, middle and inner – stories have beginnings, middles and ends.

My interest in the story of sound began when I visited

Beethoven's house in Bonn. I saw, in glass cases by his piano, the hearing trumpets that helped him compose as he lost his hearing. Shaped like ladles, they look crude to us today, but we owe a lot of music to them. They were ingeniously designed for him to attach to his head with a metal band, like headphones today, so his hands were free to compose. The end of the ear trumpet was long enough to sit level with his keyboard so he could hear the notes as he played them. These ear horns amplified and fun-nelled sound, to replicate what the ear does, for a com-poser whose world was sound and who described himself as being progressively 'exiled' from that world.

To trace the different episodes of sound's journey, I went to see Dr Ghada Al-Malky, a specialist in hearing and balance at the UCL Ear Hospital. In front of a colour-ful, large-scale model of the ear, Ghada showed me, first of all, how the poison in *Hamlet* would have travelled through the perforated eardrum and down into the king's throat. On that bright, cartoonish plastic model, it was easy to imagine the poison burning its way through the soft tissue and 'coursing through' the body. But she also showed me how we hear and how we make sense of what we're hearing.

In the history of our bodies, hearing comes long before speaking: the human ear is already fully developed around twenty weeks into pregnancy. So the newborn baby has already heard the world before it arrives into it. We hope it has heard kind, loving words whose meaning it does not understand but whose intention, tone and pitch, it senses and takes comfort from. But, of course, many tiny ears have heard threats, recriminations, insults, sobbing and shouting, long before the body they're attached to emerges blinking and sticky-eyed into the light. A foetus knows what sounds *do*, even if it doesn't yet know what they *mean*.

We can shut our eyes, but our ears are more difficult to control. From sound-muffling earplugs to £300 noise-cancelling earphones, we look for ways of counteracting our ears' unceasing activity. And even when there's nothing to hear, the ear will find something. Block your ears with your hands and all that happens is that you hear yourself hearing; you hear your body pulsing, the blood in your head, a thumping that feels at once intimate and far away.

Think of the seashell where, as children, we were told we could hear the sea – as if the shell had recorded the sea and was playing that recording over and over in its own coils and corridors. The 'shell-like ear', as the saying goes, and we mean its resemblance to a conch, a snailshell, a whelk, a periwinkle and countless other shells whose structure – inside and out – echoes our ear's. There's even a shell called 'the baby's ear', delicate and almost translucently pale. Despite the cuteness of its name, it is in fact a carnivorous sea snail.

The ear is a place, in the same way that a house or a

maze or a palace is a place, full of chambers, corridors and passages. Because part of it is outside the head and part of it inside, it is both public and private. It lets in water, rain, wind. It is vulnerable, too – think of how intrusive it feels to have a mosquito at your ear, the almost electrical shock of the insect hovering at the entrance to your head, as though the next stop is your brain. We adorn the ear with rings and studs: it can be seen, and all that can be seen can be decorated. Inside, however, the ear's work goes unseen, and as Ghada took me through it, she revealed an organ with all the complexity of the most sophisticated recording studio.

Let's take the sound we are most familiar with: our name – that strange, intermediate part of ourselves that is public (it's on our tax forms, our bank card, our pay-slips) and intimate (our parents gave it to us, it's inside us, we've grown into it). Like our ears, our names face both outwards and inwards. Let's imagine our name is called out to us across a room or in a busy street. We hear it and immediately turn to see who called us. It's a simple action. We probably never think about it – luckily, because life would be impossible if habit didn't save us from our body's exhausting surprisingness. We catch our name because we are attuned to it, and so something in our auditory habit, our hearing self, isolates it above the irrelevant stuff: street sounds, sirens, general hubbub.

Our name reaches the outer ear, the pinna, or the

auricle, as sound waves. The outer ear, the bit with the piercings and the wax and the bits of blown-around grit, that earlobe we fidget with when we're thinking, helps determine the direction the sound waves are coming from. It collects them and funnels them through the ear canal, where they cause the eardrum – the portal to the middle ear – to vibrate. It's also called the tympanic membrane, and when it's doing its job it looks like the throbbing of a loudspeaker as music is played through it.

What happens next has a beautifully mechanical simplicity. The eardrum is connected to the three smallest bones in the body: the malleus, the incus and the stapes, also known as the hammer, the anvil and the stirrup. It sounds like a smithy or a workshop, and in a way that's what it is. The malleus is connected to the eardrum, and pulls and pushes the incus, which in turn pulls and pushes the stapes, which acts as a piston creating waves in the fluid of the cochlea in the inner ear. As the eardrum moves, they move. Together, these bones, known as the ossicles, or the ossicular chain, respond to the pressure waves in the middle ear and transmit them to the inner ear. When Ghada showed me, it looked like a very basic hydraulic system, like one of those educational toys that teaches children basic mechanics. If the outer ear looks like a seashell, the inner ear is more like a snailshell. Here, the mechanical, hydraulic system of the ossicular chain gives way to the electrical. The cochlea is a coiled, spiralling tube full of fluid. It is lined with hair cells which, as they move, send an electrical impulse to the cochlea nerve which then transmits them to the brain. (We are getting deep into the head now, and on the model it all appears disconcertingly close to the brain.) The louder the sound, the more hair cells move. But the hair cells in the cochlea also distinguish different pitches for us. The hairs at the bottom, at the base of the coil, help us hear high frequencies, and those at the top, at the end of the snailshell, are responsible for the low frequencies. That is a spectrum of between 20,000 Hz and 200 Hz, ascending and descending in the same way as the keys on a piano. As the fluid in the cochlea moves, it sets off movement among the hair cells that create electrical signals that are passed by the hearing nerve to the brain, via the auditory cortex, which sorts sound waves from outside your ear into information in your brain. If any single one of these functions doesn't work, or breaks down, then our hearing is affected, with only some frequencies and pitches reaching their destination. This is why deafness and hearing loss are as complex and as varied as hearing itself, and why the image of the deaf person living in complete silence is a stereotype.

We hear our name and then we turn – how do we even know which way to turn? Because our ears do something else for us – something so basic that we don't notice it, though we would notice immediately if they stopped doing it: they contribute to our sense of equilibrium and our sense of direction. The reason we know which way to turn when we hear our name, and the reason we can stand up and balance our complex bodies in an infinity of daily tasks, is that the ears contain three loops, called semicircular canals, that sense movement and stillness and send signals to our brains. One loop senses up and down movement, another side to side movement, and the other senses tilting movement. And because our ears are on opposite sides of our head, tiny variations in the timing, volume and frequency of what they hear not only contribute to our sophisticated responses to, say, music or singing, but to our ability to locate where sound is coming from, to turn towards the person speaking to us, and to cross the road or the room to meet them.

All that's happened is that someone has called out our name and we've turned around. This story, unlike most stories, takes longer to tell than it took to happen, and I think that's a pretty good definition of the everyday miracle that is the human body at work: something that takes more time to explain than it took simply to be.

Was someone trying to catch our attention? Maybe there's someone with the same name a few metres away, at the next table, across the street? Ah yes, that's it: look, it was someone else with the same name. We switch off again, go back to chatting or reading our book or waiting for our taxi. Our ears zone it out again, but they will stay alert, they will stay open, they will never switch off.

THYROID

CHIBUNDU ONUZO

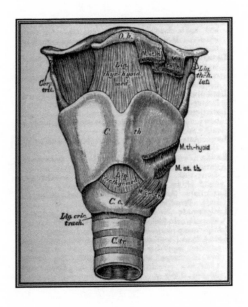

One night, while my aunt lay in bed with her husband, she rolled over the space between them and placed her head on his chest. Pressed against his ribcage, she heard his heart beating at great speed. She was pleased. Fifteen years of marriage, four children, hips widened, waist thickened and yet, she could still raise his pulse. Minutes passed with no other response from him. My uncle was fast asleep, dead to her sex appeal. Why, then, was his heart racing? It was the first sign that his thyroid was malfunctioning.

The thyroid is a bow-tie-shaped gland clipped to the base of your neck. We all have one in the same rust-red colour, nature indifferent to personal tastes. The earliest recorded accounts of the thyroid gland are in Greek medicine, by Hippocrates and Plato, who both identified the thyroid over 2,000 years ago. However, they guessed wrong about the function of the gland. They thought it was responsible for lubricating the respiratory passageways. Millennia later, European doctors still didn't know what the bow tie was for.

One mistaken theory, popular in the seventeenth century,

was that thyroids were meant to beautify a woman's neck. A long, swan-like neck was supposedly enhanced by the slight bulge of an enlarged thyroid. Perhaps this theory was developed because of the glut of Renaissance paintings that show the Madonna with a swollen neck. Da Vinci, Caravaggio, Titian, all imagined the Virgin in different poses: dandling the Messiah on her knee, teaching the Christ-child to walk, ascending to heaven in a swirl of clouds; but always, whether earthbound or floating free in the sky, Mary had this signature thickness at the base of her throat.

Did these artists know that an enlarged thyroid was the root of the swelling on their models' necks? Did the polymath Da Vinci, who sketched the thyroid after dissecting a mammal, link the gland to the swelling in his models' throats? Probably not. It is unlikely they knew that these models from Tuscany and Umbria, these young girls chosen to pose as the Madonna, had goitres, another sign that the thyroid is malfunctioning.

The thyroid gland produces a hormone called thyroxine. To make thyroxine you need iodine in your diet, and if your body doesn't have enough iodine, the thyroid overworks itself and swells into an attractive goitre (like the Madonna's), and then into a grotesque goitre, the size of a turnip.

The sea is the richest source of iodine. Seaweed, kelp, cod, all will keep your thyroid in check. Slightly pricier, especially in centuries past, rich foods like yoghurt and cheese also contain iodine. Until the twentieth century, if you lived inland and were poor, were far from the coast and penniless, then your thyroid was screwed. Then, in the early twentieth century, someone in the US had the bright idea of iodising salt. If you add iodine to salt, even the most landlocked landlubber can produce thyroxine with ease.

So what exactly does the thyroxine hormone do? It regulates the body's metabolism, the rate at which we develop and grow. Intelligence, the onset of puberty, who is top of the class, whose period came late, who is six feet tall, whose chest remained as flat as a Dover sole: all linked to the thyroid gland and the magical hormone it produces. It sounds like something out of a Lewis Carroll story. Drink this and you'll grow big and strong. Don't drink it and you'll be stunted.

The amount of thyroxine the thyroid gland produces is regulated by another gland: the pituitary gland, located at the back of the brain. The pituitary keeps everything 'Goldilocks', not too much thyroxine, not too little. Just right. But sometimes the thyroid gland goes rogue. Stress can be a trigger. If the thyroid produces too much thyroxine, you start burning up, literally.

You feel too hot, can't sit still, legs jiggling, hands trembling. You eat and eat, but you keep on losing weight. Your body incinerates the calories, your metabolism goes like lightning. Your heart speeds up. Even at rest, it beats like you've run a marathon. Even asleep, when your wife places her head on your chest, your cardiac muscle is booming.

Sometimes, you can't even sleep. Your eyes start to bulge, at first just like anyone who is sleep deprived, but then they swell further, abnormally, popping out of your head like billiard balls. Fat is building up at the base of your eyes and literally pushing them out of their sockets. Everything is rush, rush, rush. Go, go, go. Life is on speed. But you're tired and you're dying. This condition is called hyperthyroidism. If it remains untreated, you will have something called a thyroid storm. One day, your body will be unable to take the pressure of the hurricane raging inside you. You will go into cardiac arrest and die.

And then, sometimes, the thyroid falls asleep. It produces too little thyroxine. If you're born with too little thyroxine and it's not spotted early, you develop cretinism. Your height is stunted. As an adult, you'll be lucky to reach four feet. Your bones are small and weak. Puberty is seriously delayed. No eggs to menstruate. No hairs sprouting under armpits. No pimples.

But perhaps the best known effect of cretinism is the effect the condition has on intelligence. Your parents can be neurosurgeons, you can eat all the eggs and brain food you like, but if you don't have enough thyroxine in your bloodstream, your chances of learning the alphabet, let alone acing the eleven-pluses and going on to Oxbridge, are slim indeed. Thus, if you've ever called someone a cretin, you probably carried out a misdiagnosis. Cretinism is a genuine disease and the word cretin should not be used as a colourful way to describe your friends.

So the thyroid is important and you need it like Goldilocks' porridge. Not too hot. Not too cold. Just right. A famous surgeon discovered this at the expense of his patients. By the late nineteenth century, people had a vague idea what the thyroid gland was for, and they knew that sometimes, when it wasn't working properly, it made you hyper and begin to burn up. So this surgeon decided the obvious thing to do was cut out the thyroid. It's giving you trouble, get rid of it completely. Incision in the neck, bow tie snipped off, job done.

At first results were spectacular. Doctor, I can sleep now. My heart's beating fine. I'm not jumpy any more, no longer on edge, no more feeling like I'm about to launch myself off a cliff at any moment. People flocked from everywhere to have their thyroids removed. And then the doctor began to notice that something was going wrong with many of his success stories. They slowed down, which was what he wanted, and then they slowed down too much. They were lethargic, they felt cold in summer, their eyes grew puffy like people who spent most of their lives asleep. Their personalities began to fade, the sparks of intelligence stamped out, their faces expressionless, vacant, blank until they became human turnips. They needed thyroxine.

In the early days, people who'd had their thyroids removed took to eating the powdered thyroid glands of pigs, cattle and other mammals that also produce thyroxine to compensate. And then, in the 1920s, British chemists Charles Robert Harington and George Barger worked out how to synthesise the hormone. So now if your thyroid gland needs to be removed, you take your thyroxine in tablet form, with a concentration calibrated to your needs, and everything is just right.

And what if the thyroid becomes cancerous? Radiation is used to treat most cancers but a very special radiation is used to attack thyroid cancer. As mentioned earlier, the thyroid needs iodine to make thyroxine and so it's the only place in the body where iodine is stored. Patients with this rare cancer are injected with radioactive iodine, which travels straight to the thyroid and begins to attack the cancerous cells.

Now, after you've been given this radioactive treatment, for almost three weeks, you become radioactive yourself, a human torch, like a superhero straight out of a Marvel comic. Your poo is radioactive, your saliva is radioactive, your clipped toenails are radioactive, your pee, your sweat, your hair. You have to be kept in quarantine until your glow goes out. It sounds like a sci-fi movie but it's true and also the premise of *Get a Life*, one of my favourite

Nadine Gordimer novels. The protagonist undergoes the radioactive iodine treatment, is separated from his family for eighteen days and is forced to re-evaluate life.

Which is what I've been doing since I began working on this essay. I keep putting my thumb to the base of my throat and pressing down. I can feel the sinews in my neck, the fat padding my veins, but I can't feel my bow tie. All's well and good then, my thyroid is in check. But only because I was born in the twentieth century. My West African ancestors were not coastal people. They lived far inland, far from the iodine the thyroid needs. I definitely see goitres in my past – huge, pendulous ones. If a goitre grows too large, it can press down on your wind pipe, making breathing difficult. Or it might press down on your voice box, coarsening your voice, giving you a rasp. Three hundred years ago, anywhere in the world, a woman with a horrible growth on her neck, a hoarse voice and no monthly cycle, no children: a witch.

And then my mind turns to the frivolous. I have a post-Christmas waistline, a belly rounded on turkey and jollof rice. An extra dose of thyroxine to speed up my metabolism and burn some of that fat off perhaps? Can the hormone be used like that? A super dieting pill? It'd make someone billions surely. But I'd be skinny and on edge. I'd have none of the smug satisfaction that comes from being underweight these days. It turns out that I'm not the first person to have this bright idea. Google lets me know that thyroxine has been used in a weight loss pill before, but the side effects were disastrous.

Once I tell people I'm writing an essay on the thyroid, the stories start pouring in. The grandmother in the village hut with an egg-shaped goitre that hung from her neck till she died. I am reminded of a teacher of mine with eyes that bulged behind her glasses. She most likely had too much thyroxine in her blood, I realise now. We didn't know that when we called her 'frog eyes'. A friend tells me about her mother who started feeling cold and lethargic in her sixties. 'It's menopause,' the doctors told her. 'Body's natural changes. Partly in your mind.' Symptoms persisted, tests were run and her thyroid was found to be falling asleep.

Most of the stories I hear are about women. The thyroid is more likely to act up in us. Perhaps this is why the history of the gland is intertwined with vanity and not valour or intelligence. A large heart for courage … in men. A large head for intelligence … in men. A large thyroid for beauty … in women. Of course.

Or perhaps it's just because the thyroid is in a place the eye is immediately drawn to. An aunt of mine had her thyroid removed by a hack surgeon who left a penny-size scar on her throat. She took to wearing scarves, elegant silks knotted round her neck like an old-school film star. She could also have worn a thyroid choker, a necklace that fits tightly to the neck with a large pendant concealing the base of the throat. If the thyroid was on your back, or your foot, or your upper thigh, a place where a swelling would go unnoticed, there would be no paintings of abnormal thyroids, no Madonnas with goitres, no jewellery fashioned to hide the scars.

I am full of trivia by the time I finish researching this essay, enough random facts to be your phone-a-friend. Enough facts to irritate everyone around me for the next year. Look at your watch, if you're wearing one. Did you know the man who invented the second hand, Robert Graves, did so to track the fast heartbeats of his patients who were hyperthyroid? Or that eating too much kale can give you a goitre because it's one of many goitrogenic foods (like cauliflower, broccoli, turnips and radishes) that inhibit your body's ability to store iodine. So go easy on the greens.

I am also full of wonder at the end of this essay. Like most of us, I take my body for granted. I live in the most complex, intricate machine and as long as it wakes up in the morning, and goes to bed at night, I am uninterested in its inner workings. And, all the while I go about my daily business of writing and reading and thinking, there is a small furnace at the base of my throat keeping things 'Goldilocks'. Not too hot. Not too cold. Just right.

LIVER

IMTIAZ DHARKER

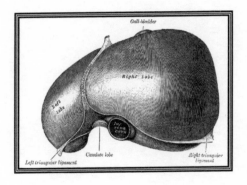

When I was small my friend Catherine and I were always in and out of each other's houses. Cathy's mother often said Cathy was her sweetheart. My mother said I was a piece of her liver.

I never thought this was odd. I knew exactly what she meant, and it felt quite natural to skip between languages with different rules. Now that I have been asked to choose an organ and write about it, I remember that the location my mother claimed for her deepest feelings was the liver. She wasn't the only one. Physicians in ancient Rome, Greece and across the Arab world also believed the liver was the true seat of love, the organ with the most fundamental role: making fresh blood rather than just pumping it around, controlling the emotions, the temperament, the character.

Poets and artists have always been quick to steal medical knowledge for their own purposes. So an Arab poet would say 'You are the soul of my soul, the blood of my liver' or 'Her look is a spear in my liver'. In the Bible, in Lamentations, Jeremiah howls, 'My liver is poured upon the earth,

because of the destruction of the daughter of my people.'
And to this day an Egyptian dancer holds her hands over
her liver to express extreme passion. So it is no wonder
that when I left home my mother said on the phone, 'My
liver has been torn apart,' and I imagined her standing
there with her hand pressed to the right side of her body.

I think of the liver's heavy lobes and its tawny satin
surface, held safe in a cage of ribs, in the upper right of the
abdomen, working as no other organ does to clean and
purify the blood, and throw bile and poisons out of the
system. For me as a poet, it is a uniquely creative, regenerative force, and I am certain this is why Pablo Neruda
singled it out for praise too, in his 'Ode to the Liver' (translated here by Heberto Morales and Will Hochman):

> … always living
> with your own dark
> filtering …
> […]
> you suck and score
> […]
> you give home
> to life's enzymes
> and grams of experience
> collecting liquors
> at this song's party
> and after cleaning up,

you are warmly last
to say goodbye.

The liver was part of everyday conversation in my house.

My mother didn't just speak about her own liver but my father's as well. If she had had the words in English she might have called someone his bosom buddy, but in Urdu she spoke of his *jigari dost*, a friend who lived in his liver. She also used the word to tell us to stand our ground, show our mettle, be courageous, *jigar me dum rakh* – keep it in the liver. For her, courage, like love, grew in the liver.

Years later I found that Shakespeare agreed with her. Macbeth says, 'Go pricke thy face, and over-red thy feare, Thou Lilly-liver'd Boy'. A coward was someone with a bloodless liver; as Sir Toby Belch says in *Twelfth Night*, 'For Andrew, if he were opened, and you find so much blood in his liver as will clog the foot of a flea, I'll eat the rest of the anatomy.'

A healthy liver showed in a ruddy complexion, a sign of robust good health. A liver that didn't function well was believed to cause mental as well as physical weakness (and you still hear people using the word 'liverish' as they get out the Andrews Liver Salts). If you called an Elizabethan liver-faced it meant they were mean-spirited. If you said they were liver-sick you meant they had dropsy, which we would call hepatitis or cirrhosis, and a gin liver would have been cirrhosis caused by alcohol.

In search of troubled and resurrected livers, I make my way to the Whittington Hospital in London to follow gastroenterologist Dr Darius Sadigh on a ward round. Before the round, the doctor discusses each patient with the discharge coordinator, psychiatrists, nurses, CMTs, first-year doctors, students, occupational therapists.

The first patient they see is Adam, whose hepatitis C has led to cirrhosis. A diabetic on insulin, he has suffered malnutrition, and his limbs are stick-thin, the stomach swollen. 'Is there any pain?' the doctor asks. 'No,' says Adam, his eyes locked on the doctor's face the whole time, watching him tap at the abdomen as if it were a door with an answer behind it.

Mel is fifty-five with a cirrhotic liver. She looks seventy-five. 'How are you feeling?' the doctor asks. 'Absolute crap,' she says. Mel is going home but, he says to her, 'When you go home, you can never drink again. If you drink again you will die. If you stay off alcohol you can do very well.' Mel is nodding but her eyes are sliding off the doctor's edges.

Jane is a tiny bird-like woman. She lives with her husband but has not spoken a word to him for ten years. They move around each other in silence and as soon as he leaves the house, she drinks.

'What do you understand about your liver?' asks the doctor.

'In trouble.' Used as she is to living in silence, her answers are terse and mostly verbless.

'Why do you drink?'

'Unhappy.'

'Is your husband abusive?'

'No.'

'Patients with cirrhosis who continue drinking don't live,' says the doctor.

'Won't touch it again.'

The doctor decides to go in all guns blazing, expecting resistance. 'You need go to rehab.'

'Okay,' she says.

Hameeda Bibi is forty-nine, has a fatty liver and hepatitis C. If it is not treated, fatty liver disease has the same effect as liver disease caused by alcohol: scarring and cirrhosis. The doctor decides to refer her for an early liver transplant.

At every point I see the doctor probing not the body as I would expect, but the language. He is listening for the clues in each patient's responses, listening like a poet for what is said and what is not said, or what is unsayable.

Tan Ai has rosy cheeks and the wide eyes of a child even though she is sixty-four years old. She has diabetes and is not eligible for a liver transplant. She is hoping to travel to Malaysia, and the doctor tries to manage her expectations. 'I can walk around without help now,' she volunteers, hopefully. The doctor advises her not to expect to get on a flight. 'I would say the prognosis is a few months.' She nods, still smiling. For a moment I think that she has not understood. Then she says, 'Good, I'd rather you be open,' and she laughs, looking radiant as a girl.

***

A friend of mine announced one day, 'I have a beautiful liver.' I wondered how she knew this. 'I had a liver scan and I heard the doctor telling the students: This is a beautiful liver. Look at that smooth shape. *And* it has a great colour.'

The miracle is that the liver has the power to regenerate, the only visceral organ that does so. Depleted to as little as 25 per cent of the original liver mass, it can grow back to its full size remarkably quickly.

When doctors speak of the liver, they often bring up the myth of Prometheus. Zeus, furious when Prometheus duped him and stole fire from the gods for human use, devised a dreadful and cunning punishment. Prometheus was chained on a mountain face. Every day an eagle would swoop down, tear open his stomach and eat his liver. Every night, the liver would regenerate, only to be devoured again the next day. The punishment was eternal, everlasting. Zeus clearly had inside knowledge of the liver's power to repair itself (although the timescale of regeneration was closer to that of rats than humans), and the Greeks saw it as the location of life, intelligence and the immortal soul.

It may have been the idea of playing God that led one consultant surgeon in Birmingham to write his initials on the surface of his patients' livers during the complex transplant operation, using an argon gas coagulator. He was convicted, found out by another surgeon who operated on one of his patients years later. It emerged that he had committed the crime more than once, and it was only because the diseased surface had changed colour to a pale yellow that the initials stood out. Perhaps what led him to do it was the idea that, somehow, he could put his name on the immortal, write on eternity. I remember lines from the poet Rumi, 'Shams Tabriz, this frantic heart / has etched your name on my liver.' When my mother died of cancer of the liver, I was told she had a snag in the liver. In my mind I could see this snag, this pain knotted in to the liver, and it seemed to spell out my name.

In ancient Greece and Rome, and in Africa, it was common practice to look to the surface of the liver for signs and portents. In the Bible, the prophet Ezekiel says the king of Babylon looked for guidance before attacking Jerusalem. 'He consulted with images, he looked in the liver.' Plato, too, believed that the rational soul in the brain projected images onto the shiny surface of the liver.

As the Prometheus myth suggests, the Greeks certainly knew of the liver's power to regenerate more than 2,000 years ago, knowledge that may have come from seeing animal livers repair themselves. But it was not until 1931 that Higgins and Anderson were able to prove that a rat liver begins to regenerate hours after partial removal; it was 1963 before Dr Thomas Starzl attempted the first human liver transplant (unsuccessfully, as the patient died of bleeding during the operation); and not until 1967 that he was able to keep a child alive for over a year after a transplant. Sir Roy Calne's introduction of the immunosuppressant ciclosporin improved patient outcomes, and today, for patients whose livers have been scarred or damaged beyond the point of self-repair, thousands of liver transplants are performed across the world. However the shortage of bequeathed livers has also led to an illicit organ trade from living donors in poorer countries, and also to legal transplants from living donors.

It is quite common now for a parent to donate a piece of the liver for their child. The child, in an echo of my mother's words, becomes a piece of the parent's liver. In a living transplant, the right lobe, 70 per cent of the liver, is taken from a healthy living donor. The donor's remaining left lobe could grow back into a fully functioning liver with two lobes within six weeks. The recipient's 70 per cent regenerates too.

There was a recent news story of two girls, one seventeen years old and the other just eleven months old, on the transplant list. The bequest of a whole liver allowed the doctors to give 70 per cent to the teenager and 30 per cent to the toddler. Having shared a liver, there is every chance that each of the girls will grow her own healthy liver.

The footballer George Best famously joked, 'I spent a lot of money on booze, birds and fast cars. The rest I just squandered.' He had a liver transplant (courtesy of the National Health Service) but did not stop drinking, and died three years later. According to some doctors, news of this caused a drop in liver donations. Nowadays, to be eligible to receive a liver transplant, there has to be confidence that the recipient will try to keep away from alcohol and stay sober. The liver is seen as a gift, not to be squandered.

The liver is also a precious food, rich in nutrients for animals and humans. Orcas disembowel sharks and seals just for their liver, which is full of squalene that helps produce steroids and hormones. Nubians enjoy a great dish of raw camel liver. The French force-feed a goose until the fatty liver swells to many times its normal size to make foie gras. (And who could forget Dr Hannibal Lecter's meal of liver with fava beans and a nice Chianti?) Down the road from where I live, in Smithfield Meat Market, chefs haggle over succulent calf's liver. On the streets of Bombay's Bhendi Bazaar, mutton liver is chopped, tossed on a sizzling tava with chillies and cumin, and sold for ten rupees. People across the world demand liver from every kind of fish and animal, and they eat it sliced and fried with onions, in terrines or pâtés, in dumplings, meat pies and pasties, or stuffed into wurst and haggis.

Once a week, on the family doctor's orders, my mother used to hold my nose and pour a foul cod liver oil down my throat for the vitamins A and D. When I stopped gagging and finally agreed to forgive her with a hug, she would sigh, 'You have cooled my liver', and I'm sure that is exactly where she felt whatever it was that she felt: right there, in the centre of her being, in the liver.

WOMB

THOMAS LYNCH

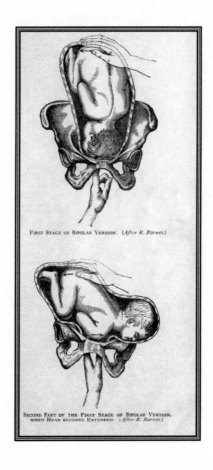

First Stage of Bipolar Version. (*After R. Barnes.*)

Second Part of the First Stage of Bipolar Version, when Head becomes Extended. (*After R. Barnes.*)

The contemplation of the womb, like staring into the starlit heavens, fills me with imaginings of Something-ness or Nothingness. It was ever thus. If space is the final frontier, the womb is the first one – that place where, to borrow Wallace Stevens' phrase, the idea of the thing becomes the thing itself. It is the tabernacle of our expectations, the seedbed and safe harbour whence we launch, first home and habitat, the garden of delights' denouement. A place where the temps are set, the rent is easy, the food is good and we aren't bothered by telephone or tax man. That space we are born out of, into the world, where the soft iambics of our mother's heart become the first sure verses of our being.

When I first beheld, as a student in mortuary school, Plates 60 and 61 in Book Five of *De Humani Corporis Fabrica – The Fabric of the Human Body –* by the great sixteenth-century physician and anatomist, Andreas Vesalius, I was smitten with ontological and existential awe. Vesalius was a disciple of the first-century Greek philosopher and medico Galen of Pergamon, who was enthused by both the rationalist and empiricist sects of medical research, and one sees in the Belgian's handiwork the male gaze on female parts he examines at autopsy and vivisection. There is almost a tenderness in the splayed cavities and skinned breast of the headless woman of his scrutiny, such precision to his illustrations.

By then I'd had a rudimentary acquaintance with the bodies of women. I knew what to touch and rub, hold and savour. But the frank exposure of the human fabric that Vesalius' images made plain were wondrous to me, unveiling as they do, apocalyptically, the beauty of both form and function. Had I not found his drawings so sumptuously instructive – corresponding as they did to the focus of my own gobsmacked gaze – I might have considered then what I consider now, here in my age and anecdotage, to wit, the notion that, though each gender has its own specific parts to play in our species' drama of 'reproduction', such issues are neither solely male nor female. Rather, they are human in scope and nature, requiring in both meaning and performance, the two it always takes to tango. We are, it turns out, in this together.

Still, it is impossible to behold a woman's body without gratitude and awe. Likewise I am often chuffed – a word which means both one thing and its opposite – by the sense that such encounters invariably affirm that we are all, in fact, the same but different. The anatomists' renderings of our private parts shows the male member is nothing so much as a vagina turned inside out, so that the adventitia, smooth muscle and mucosa of the latter reflects, actually, the phallic urgency of the former, almost as if they were made for each other, bespoke, custom fit – like sword to scabbard, hand to glove, preacher to pulpit or corpse to opened ground.

And, lest anyone assume the sword more salient than the scabbard, consider science, that great leveler. In utero, we all start as female, or, for the sensitive, gender-neutral, and only the random happenstance of the Y chromosome and its attendant hormones, six weeks into the mix, makes some of us male. Still, testes are unambiguously fallen ovaries, and the scrotal raphe a labial scar from the fusion of one's formerly female lips. The penis is a clitoris writ large, the nipples, sans lactation, ornaments to remind men of the truth that they are mostly boobs that do not work. So, whether penetration, ejaculation, ovulation, uteran contraction, fertilisation or gestation seals the reproductive deal, each is essential to this essential mystery. We are in this together; we are brought into being by the fervent collaboration of both male and female. Science provides stand-ins for the stallion and sire. 'They bring the bull's semen in a suitcase, now' my cousin Nora, in West Clare, informed me years ago, speaking of her small troupe of milking Friesians, which had a withering, possibly salutary effect upon my rampant mannishness. Men are easily made redundant, but female mammals still do the heavy bearing. Far from the second, weaker sex, the female seems the first and fiercest, like poetry to language, the one without which nothing happens.

I went to fetch stillborn babies, as a boy. Well not a boy, exactly, but not yet a man. My apprenticeship to my father's business meant I'd go to hospitals to get the tiny lifeless bodies, transported in small black boxes, such as one might tote shoes or keep one's tools. I'd return them to the funeral home: wee incubates in various stages of incompleteness and becoming. Sometimes they were so perfectly formed in miniature that they seemed like tiny icons of humanity, their toes and fingers, nose and eyes, their little selves too small, too still, but otherwise perfectly shaped and made. As with Galen and Vesalius, as with Wallace Stevens, the thing itself outweighs the idea of the thing. Thus these little fetal things, stillborn or born but not quite viable, were freighted with a gravitas, fraught with sadness, laden with a desolation born of dashed hopes and grave-bound humanity. The body, the incarnate thing, is critical to our understanding. The rationalist and empiricist sects are always some at odds. If whence and whither are the questions posed by cradle and coffin, the uterus is wellspring, headwater, home ground of our being.

In time, I'd learn to sit with the families of dead foetuses, dead toddlers, dead teenagers – the parents who'd outlived the ones they'd made, the fathers who remembered the night of bliss they'd had, the kissing and embraces, the mothers who recalled their first intuitions of gravity, their gravidness, its gravitas, the grave consequences of impregnation – an ill-at-easeness in their tummy, tenderness in the breast, the momentary hot flash of a future changed or changing utterly.

'It's only been maybe a hundred years,' says my young assistant, halfway through her childbearing years, 'that women have actually owned their uteri.' And even now, she adds, the agency of men – of husbands and fathers, bishops and politicos, no less moguls and marketeers – have too much of a say in what goes on in the hidden places of a woman's body, the womb and its attendant, adjacent parts: cervix, ovaries, fallopian tubes, the adventitia, clitoris, labia major and minor, the mons pubis, all of which conspire, as it were, to raise a chorus of praise to the mighty nature, whereby we renew, repeat, reproduce and replicate ourselves.

In the panel *Eve Tempted by the Serpent* by Defendente Ferrari, who was painting in Turin whilst Vesalius was dissecting in Padua, the pale-skinned, naked teenager's mons Veneris is obscured by the filigree leaf frond of the sapling she is plucking an apple from – the tree of knowledge of good and evil. The leering, bearded, lecherous, old-mannish-faced snake slithering up the adjacent tree is hissing temptation in her ear. It is the last moment of Paradise; the girl is in her girlish innocence, oblivious to the ramifications. Her genitalia, her tiny breasts, her consort's parts are not yet shameful. Time will eventually blame everything on her: the Fall of Man, the pain of child-birth, the provocations of her irrepressible beauty, death itself. But for now, God is still happy with creation. He has looked about and seen that it was good. It's all written down in Genesis 3. The accompanying diptych panel with Adam, perhaps erect and prelapsarian, has been lost to the centuries, so we do not see how happy he is, how willing and ready and grateful he is for her succour and company, her constancy.

On a drizzling morning in the winter of 1882, in Washington DC, a retinue of black-clad pilgrims gath-ered around a small grave in the Congressional Cem-etery to bury little Harry Miller, a toddling boy who had succumbed to that season's contagion of diphtheria. The small coffin rested on the ropes and boards over the open ground while the mother's sobs worked their way to a cre-scendo. The undertaker nodded to the man at the head of the grave to begin. He shook his head. The mother's animal sobs continued. She was bent over, like someone stabbed, wrapping her small arms round her uncorset-ted middle, holding herself together by dint of will at the point in her body where she felt the blade of her bereave-ment most keenly. People shuffled their feet in the cold, discomforted by her grief.

'Does Mrs Miller desire it?' the speaker asked. The dead

boy's father nodded his assent. The dead boy's mother quieted, though still writhing in pain at her core.

The officiant on the day, Robert Green Ingersoll, was no pastor or parson, cleric or priest. He was, rather, the most notorious disbeliever of his time, his age's Christopher Hitchens, Richard Dawkins or Bill Maher. Though stridently unchurched, Ingersoll was the youngest son of a Congregationalist minister who preached his abolitionist views, and had, as a consequence, been given his walking papers by congregants around the east and Midwest. Robert spent most of his youth shifting from church to church because of his father's politics. Because of his father's mistreatment at the hands of Congregationalists, Robert turned first on Calvinism and then on Christianity and, by the time he stepped to the head of the grave on that rainy morning in Washington DC, he was the best-known infidel in America – an orator and lecturer who had travelled the country upholding humanism, 'free thinking and honest talk', and making goats of religionists and their ecclesiastical up-lines.

'Preaching to bishops,' a priest of an acquaintance once told me, 'is like farting at skunks.' I wonder now if he wasn't quoting Robert Green Ingersoll. He taught law and lectured on Shakespeare, Reconstruction and Religious Hucksterism, and was held in such high regard by Walt Whitman that Ingersoll gave the eulogy at the great poet's funeral. He was able, it seems, to rise to all occasions.

As he stepped to the head of the Miller boy's burial site, Ingersoll began his oration:

> I know how vain it is to gild a grief with words, and yet I wish to take from every grave its fear. From the wondrous tree of life the buds and blossoms fall with ripened fruit, and in the common bed of earth, patriarchs and babes sleep side by side.
>
> Every cradle asks us 'Whence?' and every coffin 'Whither?'
>
> They who stand with breaking hearts around this little grave, need have no fear. The larger and the nobler faith in all that is, and is to be, tells us that death, even at its worst, is only perfect rest.
>
> We have no fear. We are all children of the same mother, and the same fate awaits us all. We, too, have our religion, and it is this: Help for the living – Hope for the dead.[1]

Every cradle asks us whence indeed, and every coffin whither. The abyss we consign our dead to – opened ground or fire, pond or sea or air – is incubation of a sort our sacred texts make faith claims for, hoping they are like the space of the womb, pear shaped sometimes, no more than centimetres, hormonally engaged, impregnated by mighty nature, a primal station in the journey of our being.

What bent the dead boy's mother over was the grief, felt

most keenly in her most hidden places, the good earthen, opened seedbed of her uterus, vacated with pushing and with pain, and vanquished utterly by her child's death. It is the desolation Eve must have felt when one of her sons killed the other. And the wonder Andreas Vesalius beheld when looking into the bloody entrails of the Paduan girl who first unveiled for him the mystery of our coming into being. It is the sound and sense we humans get, examining our lexicon, that 'grave' and 'gravid' share their page and etymology, no less 'gravitas' and 'gravity', 'grace' and 'gratitude'. And that the surest human rhymes of all are 'womb' and 'tomb.'

Notes

1. 'At a Child's Grave,' *The Works of Robert G. Ingersoll*, Clinton P. Farrell, Editor, p.399

PICTURE SOURCES

p. 123 Lungs, Fig. 970, *Anatomy of the Human Body*, Henry Gray, 1918

p. 135 Ears – source unknown

p. 147 Larynx, front view featuring thyroid, *A text-book of human physiology*, 1904

p. 159 The superior surface of the liver, *Anatomy of the Human Body*, Henry Gray, 1918

p. 171 *An illustrated dictionary of medicine, biology and allied sciences*, George M. Gould, 1900

ABOUT THE AUTHORS

Naomi Alderman is a British author, novelist and game designer. Her novel *The Power* won the Baileys Women's Prize for Fiction in 2017.

Ned Beauman is a novelist and journalist. His books include *The Teleportation Accident* (2012), which was longlisted for the Man Booker Prize, and *Madness is Better Than Defeat* (2017).

Kayo Chingonyi is a poet and the author of two pamphlets, *Some Bright Elegance* (2012) and *The Colour of James Brown's Scream* (2016). His first full-length collection, *Kumukanda*, was published in 2017.

Abi Curtis is a novelist, poet and Professor of Creative Writing at York St John University. Her debut novel is *Water & Glass* (2017) and her poetry collection *The Glass Delusion* (2012) won the Somerset Maugham Award.

Imtiaz Dharker is a poet, artist and documentary filmmaker. Awarded the Queen's Gold Medal for Poetry in

2014, she has six collections including *Over the Moon* and the latest, *Luck Is the Hook*.

William Fiennes is the author of novels *The Music Room* (2009) and *The Snow Geese* (2002), which won the Somerset Maugham Award. He was diagnosed with Crohn's disease aged 19.

Annie Freud is a poet whose collections have won the Glen Dimplex New Writers' Award (Poetry) and been shortlisted for the T. S. Eliot Prize. She is one of the Next Generation Poets, as announced by the Poetry Book Society in 2014.

A. L. Kennedy is a Scottish writer of novels, short stories and non-fiction, academic and stand-up comedian. Her novels include *Day* (2007), which won the Costa Book of the Year Award, and *Serious Sweet* (2016), which was longlisted for the Man Booker Prize.

Philip Kerr was the author of forty books, including the bestselling Bernie Gunther thrillers set in Nazi-era Berlin and the young-adult series *Children of the Lamp*.

Thomas Lynch is a poet, essayist and undertaker. His books include *The Undertaking* (1997), which won the American Book Award. He has been the funeral director in Milford, Michigan since 1974.

Patrick McGuinness is a British academic, critic, novelist and poet. His first novel, *The Last Hundred Days* (2011), was longlisted for the Man Booker Prize and shortlisted for the Costa First Novel Award. His latest novel, *Throw Me to the Wolves*, will be published in 2019. He is Professor of French and Comparative Literature at the University of Oxford.

Daljit Nagra was the first Poet in Residence for BBC Radio 4. His collection *Look We Have Coming to Dover!* won the Forward Poetry Prize in 2007.

Chibundu Onuzo is a Nigerian novelist whose books include *The Spider King's Daughter* (2012), which was shortlisted for the Dylan Thomas Prize and the Commonwealth Book Prize, and *Welcome to Lagos* (2017).

Christina Patterson is a writer, broadcaster and columnist. She writes, for the *Guardian* and the *Sunday Times*, about society, culture, politics, books and the arts. She is the author of *The Art of Not Falling Apart* (2018).

Mark Ravenhill is a British playwright, librettist, actor and journalist. His plays include *Shopping and Fucking* (1996) and *Mother Clap's Molly House* (2001).